C0-ALK-840

The Ecology of Vertebrate Olfaction

The Ecology of Vertebrate Olfaction

D. Michael Stoddart

Department of Zoology,
University of London King's College

Chapman and Hall
LONDON AND NEW YORK

First published 1980
by Chapman and Hall Ltd
11 New Fetter Lane London EC4P 4EE

Published in the USA by
Chapman and Hall
in association with Methuen, Inc.
733 3rd Avenue, New York, NY 10017

Printed and bound in the
United States of America

ISBN 0 412 21820 8

British Library Cataloguing in Publication Data

Stoddart, D Michael
The ecology of vertebrate olfaction.
1. Smell 2. Vertebrates – Behavior
3. Animal ecology
I. Title
596′.01′826 QP458 79–40791

ISBN 0–412–21820–8

Contents

for Brigitte, Emily and Hannah

In future research, a most critical area will be the physiological mechanisms . . . peripheral encoding . . . central processes etc. In addition to the molecular approach I think we are in substantial need of an ecologically based attempt to put the whole organism in its olfactory perspective. The world is full of as much chemosensory variety as it is of visual variety . . . we must go out into the real world and attempt to create an ecological chemistry . . . Experienced woodsmen can describe the forest they are in quite accurately when blindfolded; salmon, it appears, can recognise the chemical *Gestalt* of the water in which they were hatched, and snakes possess similar abilities without learning at all.

Gordon M. Burghardt 1970

Preface

Scientists not infrequently succumb to the frustration they feel when they have to garner often quite fundamental information about an undeveloped field from scattered publications covering many disciplines by writing their own review of the field in question. This is an invaluable exercise, particularly for those in the business of stimulating students to grapple with unfamiliar ideas and concepts, since it makes the introduction to that literature much less painful. To some extent I, too, have succumbed to this frustration by writing this book, but I have also, much more importantly, tried to develop out of this literature an olfactory perspective of the whole organism in its environment – in its feeding relations, reproductive biology, ecological isolation, social organization, ability to give warning and defend itself, and ability to navigate when displaced from home.

One event more than any other acted as a catalyst to encourage me to start this task. One evening in the arid Australian bush, as I was returning to camp, the stirring of the air bathed me in a host of smells I had been unaware of in the stifling heat of the day. I found their effect on me quite extraordinary, for they refreshed and revitalized me more than I would ever have imagined possible. I had a rare glimpse of what it must be like to be macrosmatic – to rely on one's nose for one's sensory input. I owe a debt of gratitude to the Zoology Department of Monash University, Melbourne, for providing the facilities and intellectual atmosphere which enabled me to formulate the basic shape of this book, and to the research analysts of the University Library who, by their skilful use of the Australian Medlars Information Retrieval Service, provided me with much needed background literature.

I acknowledge gratefully and freely the help I have received from a number of colleagues, in particular Eduard Zinkevich for help with the details of Table 2.3, and Gillian Sales for useful criticism of much of the text. Their help has been invaluable, but I, alone, take full responsibility for the errors and omissions which remain. Many colleagues have been kind enough to allow me to reproduce photographs from their original negatives and I extend my grateful thanks to them. Where these have been published before, I acknowledge the permission granted me by the publishers for their reproduction here. The following people have helped me in this way:

Dr F. Bojsen-Møller, Copenhagen, Denmark, Fig. 1.12
Dr J. G. Ehrenfeld, New Brunswick, USA, Fig. 2.1 (c)
Dr S. Erlinge, Lund, Sweden, Fig. 6.7
Dr R. Liversidge, Kimberley, South Africa, Fig. 7.2
Dr G. McKay, Sydney, Australia, Fig. 4.11
Dr D. Müller-Schwarze, Syracuse, USA, Fig. 2.5
Dr B. P. M. Menco, Utrecht, Holland, Fig. 1.6
Professor J. D. Pye, London, Fig. 1.14 (a)

All remaining photographs were taken for me by Mr A. Howard, Zoology Department, King's College, University of London, with the exception of Fig. 2.1 (d, e and f) and Fig. 7.4 which were taken by myself. The material for Fig. 7.4 was prepared for scanning electron microscopy by the University of London Stereoscan Unit under the direction of Lyn Rolfe. Geoffrey Tomkins generously allowed me to use his unpublished data in Fig. 2.4, and Dr R. Liversidge kindly sent me the springbok hairs from which Fig. 7.4 (d) was made.

The following text figures and tables have been reproduced by kind permission of the author(s) and publisher concerned:
Fig 1.1 (Indiana University Press), 1.2 (Zoological Society of London), 1.5 A (Edward Arnold Ltd.), 1.5 B, C and D (Plenum Publishing Corporation), 1.10, 1.11 and 1.12 (Cambridge University Press), 2.5 and 2.6 (Academic Press), 3.6 (Reprinted from "The aggressive behavior of *Storeria dekayi* and other snakes with especial reference to the sense organs involved." by G. K. Noble and H. J. Clausen in *Ecological Monographs*, **6**, 301, by permission of Duke University Press. Copyright 1936 by the Ecological Society of America) 3.7 (Copyright 1975 by the American Psychological Association, Reprinted by permission). Table 4.3 (The Veterinary Record), Fig 4.5 (Zoological Society of London), 4.6 (Plenum Publishing Corporation), 4.8 (The Wistar Institute), 4.12 (University of Chicago Press), 4.13 (Academic Press), 5.5 (Copyright 1972 by the American Psychological Association. Reprinted by permission.) Tables 5.1 and 5.2 (Royal Physical Society of Edinburgh), Fig 5.4 (Royal Physical Society of Edinburgh), 6.2 (Commonwealth Scientific and Industrial Research Organisation), 7.6 (Alfred A Knopf Inc.), 7.7 and 8.3 (Macmillan Journals Ltd.), Table 9.1 (Plenum Publishing Corporation).

Finally I owe more than just thanks to my wife and daughters for their forbearance over the past eighteen months. Without that I could have achieved nothing.

London, March 1979 Michael Stoddart

Introduction

The power of odours to capture man's mind and stimulate his body has been recognized for as long as man has been aware of his ability to control his own destiny. During the Argonautic expedition, Jason and his crew put into the island of Lemnos, leaving Heracles alone on board. The ladies of Lemnos burnt fragrant substances on the altar of Aphrodite and the air was heady with aroma. The tired sailors, most agreeably bemused by the aphrodisiac smoke, would most likely have abandoned any further thoughts of Colchis and the Golden Fleece had not Heracles, angry at the delay and out of reach of the smoke, stormed onto the island and forcibly marched Jason and his men back on board. Yet, perhaps surprisingly, this power, which is far more keenly developed in vertebrates other than man, has never been extensively researched and the findings exploited for man's use. Anatomists and physiologists throughout the duration of the development of medical and biological science have sought earnestly to describe the mechanisms of sensory stimulation, but olfaction has lagged well behind the other senses. This is probably because human beings living in a civilized world think they are not particularly disadvantaged by a defective nose and so, quite understandably, a disproportionate amount of investigative effort has been directed away from the sense of smell and on to the senses of sight and hearing. It is probably also because it was not until quite recently that analytical apparatus of the required sensitivity became widely available and the timing of its availability happily coincided with a mushrooming of interest in the behavioural and ecological sciences. Anatomical and physiological studies of the most intense kind are currently in progress in many centres around the world, and one may confidently expect that soon the final links in our understanding of the chain of events leading from the first encounter of the odorant with the nasal membranes to the animal's appropriate response will be forged.

In our quest for the mechanisms of sensory arousal, we must not lose sight of the need to view the olfactory system as part of an integrated perceptual organ, the systems of which accommodate to ever-changing levels of stimulation, yet allow the animal to make a constant perception about its environment. A spirited attempt to view the combined sensory systems as a perceptual organ is contained in James Gibson's enlightening book entitled *The Senses Considered as Perceptive Systems*. Written, in his own words, 'twice – once in 1958–59 and again in 1963–64', the book was a little too early to catch the literature on how olfaction aids and assists animals to cope with their environment, which started its vigorous growth in the mid-1960s. Gibson's book is surprisingly little known to students of behavioural ecology, but its value is such that it is well worth reproducing here a few short extracts from its introduction.

> It can be shown that the easily measured variables of stimulus energy, the intensity of light, sound, odor, and touch, for example, vary from place to place and from time to time as the individual goes about his business in the environment. The stimulation of receptors and the presumed sensations, therefore, are variable and changing in the extreme, unless they are experimentally controlled in a laboratory. The unanswered question of sense perception is how an observer, animal or human, can obtain constant perceptions in everyday life on the basis of these continually changing sensations.

> The active observer gets invariant perceptions despite varying sensations. He perceives a constant object by vision despite changing sensations of light; he perceives a constant object by feel despite changing sensations of pressure; he perceives the same source of sound despite changing sensations of loudness in his ears. The hypothesis is that constant perception depends on the ability of the individual to detect the invariants, and that he ordinarily pays no attention whatever to the flux of changing sensations.

> [The] five perceptual systems overlap one another; they are not mutually exclusive. They often focus on the same information – that is, the same information can be picked up by a combination of perceptual systems working together as well as by one perceptual system working alone. The eyes, ears, nose, mouth, and skin can orient, explore, and investigate. When thus active they are neither passive senses nor channels of sensory quality, but ways of paying attention to whatever is constant in the changing stimulation. . . . The movements of the eyes, the mouth, and the hands, in fact, seem to keep changing the input at the receptive level, the input of sensation, just so as to isolate over time the invariants of the input at the level at the perceptual system.

> The perceptual systems, including the nerve centers at various levels up to the

brain, are ways of seeking and extracting information about the environment from the flowing array of ambient energy.

Now that we are at long last recognizing the manifold ways in which odours control and guide the behaviour of animals and some of their physiological processes, we should attempt an overview of olfaction to parallel Gibson's ecological view of optics. I do not dare presume that this book fills the gaps in Gibson's scholarly work, but it does tackle its subject matter from the same viewpoint, i.e., what is actually perceived by the animal. Perhaps I am premature with my efforts. Certainly the number of discoveries that could be rated as fundamental to the understanding of olfactory processes and their involvement in control of biological activities shows no sign of decreasing, but probably enough chips of the mosaic are available to allow a reasonable attempt to be made at describing not only the shape but also a little of the quality of the interlocking patterns of the pavement. In adopting Gibson's basic viewpoint, I have reduced to an indispensable minimum descriptions of the anatomy of olfactory structures and the physiology of their systems. In any case, the student of whole-animal biology would find many aspects of these matters make for turgid and tedious reading.

I have tried to review the biology of olfaction from an identifiably ecological viewpoint, for to do otherwise would miss the point about the perceptual quality of the olfactory system. The adoption of this viewpoint has a hidden intrinsic advantage which I have tried to reveal in the final chapter. In a world facing problems of food production on a scale quite unimaginable to our forefathers, scientists are being asked to control ever more pest species, to enhance further the productivity of domestic stock, and to exploit natural populations at greater advantage to mankind. Recent experience with pesticides and chemicals designed to increase growth rates in cattle has been bad, with the inevitable result that the need for radical approaches to the world's ecological problems, which will not further weaken or destroy its brittle frame, are greater now than ever before. The need will increase. Concepts are emerging which must be refined and developed, but it is already clear that students of olfactory biology have much to offer. This is hardly unexpected since olfactory perception penetrates deeply into every aspect of the lives of animals, and odours constitute an omnipresent part of their environment. In the face of this incomparable environmental challenge, the entire weight of scientific knowledge and endeavour must be brought together and beamed onto the problems affecting our very survival. This book may draw the attention of front-line researchers to the potential of artificial interference with the olfactory environment which is part of the lives of vertebrates, and thus help, in a small way, to focus the beam. For readers less involved with ecological matters and who wish only to know a little about how olfaction works and how it is used, the book will, I hope, prove to be a useful digest of its main components.

1

The olfactory system of vertebrates

The ability possessed by animals to be able to assess the quality of their surroundings is as old as life itself and is fundamental to the whole of evolutionary and ecological theory. Decisions taken by an individual about such things as what food it should eat and which mate or nest site it should select are based on an analysis of perceptual input received from all its sensory systems. These systems are not equally developed in all taxa; nevertheless perception of the chemical quality of the environment is a basic necessity to all animals irrespective of their cellular composition and evolutionary position. The basic nature of chemical perception is demonstrated by the occurrence of a chemical sense in every species of vertebrate, while some have secondarily lost the ability to perceive light (e.g., *Amblyopsis* (Pisces); *Proteus* (Amphibia); *Typhlotriton*, *Typhlops* (Reptilia) and *Bathyergus* (Mammalia)) and sound (e.g., many snakes and burrowing lizards). Because systems and structures are lost during the course of evolutionary time in response to environmental changes, it follows that universal retention of chemical perception in the face of those changes indicates the overwhelming and widespread usefulness of the ability.

The universality of olfaction in the vertebrates is reflected in the anatomical simplicity of its system. In stark contrast to the anatomical position of visual and acoustic receptor cells, which are guarded by membranes, fluid baths, bones and other structures serving to transmit and attenuate the signal waves before they are perceived, olfactory receptor cells boldly protrude into the environment, unshielded save for a thin ever-changing veil of mucus. The first structure encountered by an inspired odorant molecule is an exposed sensory cell that is no less than a forwards projection of the olfactory centre of the brain. The structure of the gustatory, or taste, system is very similar and equally universal in its occurrence. While the difference between the sense of smell and taste is clear in terrestrial, air-breathing vertebrates, it is less clear in aquatic forms for which all chemical stimulation derives from molecules dissolved in water. That two functional systems exist in fishes is suggested, if by *a posteriori* reasoning, by the presence of two well-defined anatomical structures, one associated with the nostrils (though divorced from the respiratory system) and the other with the mouth and sensory barbels. It is further suggested by experiments, performed many decades ago, which revealed that sharks hunt with their nostrils, but accept or reject prey according to its taste. Innervation of the two structural systems is also quite distinct, so it appears reasonable to assume that the systems are as functionally distinct in fishes as they are in terrestrial forms.

The olfactory system is triggered when a stream of molecules from a source enters the nasal cavities and makes contact with the sensory epithelium. This is exactly comparable to the manner in which the visual and acoustic systems are triggered. Leaving aside the question of *why* certain odours smell as they do, the olfactory receptors are every bit as sensitive to change in the quality of the incoming signal as are the visual and acoustic receptors. Thus a complex stream of messages can be received and analysed. Each sensory system possessed by vertebrates is specialized to be maximally effective under a restricted, and different, set of environmental conditions.

For an object to be perceived visually it is necessary that there be enough reflected light available to transmit the image to the perceiver and that no solid structures obscure its path. The presence of light is unnecessary for a sound to be heard, so sounds are particularly useful for communication at night. But as sounds are easily distorted and deflected by structures in their paths, their ecological usefulness is somewhat restricted to open environments where physical obstructions are few. For the perception of an odour, however, it is necessary only for the perceiver to be close enough to the odour source so that some scented molecules can be swept up into its nose. The overwhelming advantage of olfactory signalling over visual and acoustic signalling is that the signaller need not be present at the site of odour emanation – all that is required is the presence of some molecules which could have been left behind by the signaller at some earlier time. As far as food finding is concerned, this is a distinct disadvantage, unless the perceptive system is able to cope with a concentration gradient caused by a temporally induced decay, but it is highly advantageous in most social situations (Chapters 4, 5 and 6). A further point is that environmental conditions exert a marked influence on the speed and accuracy of translation of all sensory signals. Odours are particularly influenced by the movement of air, for they depend for their transmission upon Graham's laws of gaseous diffusion and this reduces their usefulness in certain situations. Intrapopulation signals which must be transmitted rapidly are very seldom, but can be, olfactory, although they may more frequently be visual and/or acoustic signals reinforced with olfactory cues.

1.1 Anatomy

The anatomy of the olfactory apparatus is remarkably similar throughout the class Vertebrata. There is an inlet to, and an outlet from, a chamber in which is held a thin sensory membrane of sometimes immense area. In the terrestrial vertebrates the outlet from the chamber enters the respiratory tract, posterior to the soft palate, but in fishes both inlet and outlet are positioned on top of the head and there is no contact between the olfactory and respiratory systems in any species except the lung fishes. The sensory membrane, or olfactory epithelium, is arranged in a variety of ways so as to present the maximum surface area to the inspired molecule stream, and the anatomy of the external nares ensures that the stream enters at the correct volume and direction. In fish, the epithelium lies on a series of bony or cartilaginous lamellae which altogether resembles a rosette or the laminar plates of a coral (Fig. 1.1) (Kleerekoper, 1969). During the life and

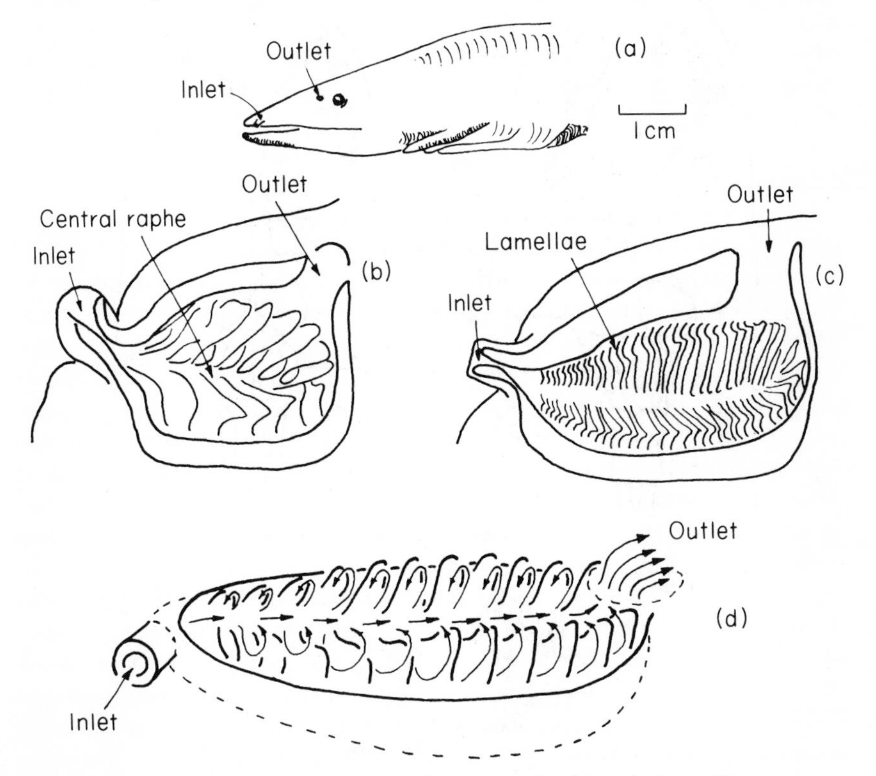

Figure 1.1 Diagrammatic view of the olfactory capsule of the eel, *A. anguilla*.
(a) Head of adult eel – drawn from life.
(b) and (c) Internal structure of the olfactory capsule of a young and old specimen respectively. (Redrawn from Kleerekoper, 1969.)
(d) Diagram of water flow over the lamellae.
(Redrawn from Kleerekoper, 1969.)

growth of a fish new lamellae develop anteriorly and proceed to enlarge laterally. Within a species, the number of lamellae present, and their size, depend upon the age of the individual, so the amount of olfactory epithelium present increases throughout life.

A remarkable feature of the biology of the bathypelagic fish, particularly the deep sea angler fish (Ceratiodae) and Gonostomatidae, is a well-marked sexual dimorphism of the olfactory organs (Marshall, 1967). It occurs in 80% of all bathypelagic species. This is not true of the meso- and benthopelagic fishes which show no marked sexual dimorphism of olfactory organs. Both groups have well-developed organs however, but females not as well-developed as males. The males of deep sea anglers are very much smaller than the females and associate

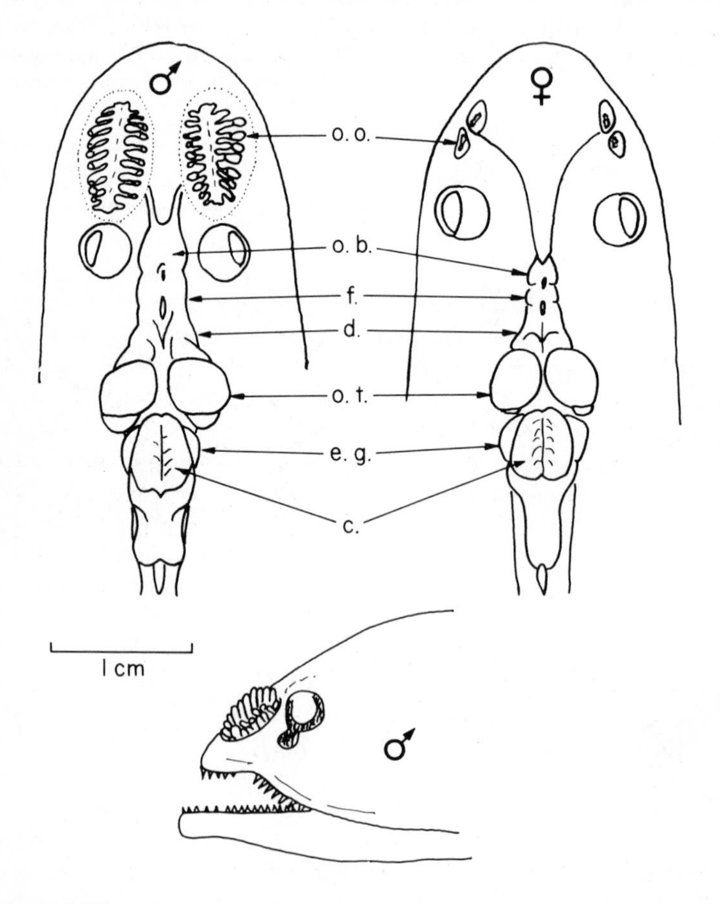

Figure 1.2 Olfactory organs and brain of the bathypelagic fish, *Cyclothone microdon*. c. cerebellum; d. diencephalon; e.g. eminentia granularis; f. forebrain; o.b. olfactory bulb; o.o. olfactory organ; o.t. optic tectum. (Redrawn from Marshall, 1967.)

with the females in a parasitic relationship; in other deep sea fishes the males are free-living, but are much smaller than the females. In these depths, from 1000–4000 m, low biological productivity dictates low populations of carnivorous fishes and a correlated low chance for the sexes to meet. Selective pressures ensuring that, once contact between sexes has been made, male and female never part have a high survival value as the numerical success of the ceratiods shows. Relatively little is known of the developmental biology of the bathypelagic fishes, but it appears that the olfactory organs of the males start to increase in size at the onset of metamorphosis from the pelagic larval stage. By the time the adolescent fish is in the region of its potential mate, the disparity between his and her olfactory organs is very great (Fig. 1.2). Although experimental proof is lacking, it seems likely that the male is attracted to the female by following a gradient of odour emanating from her. Even at depths of 2000 m water currents of up to 15 cm/s have been recorded and, although zero currents are sometimes encountered, a mean of 5 cm/s is quite normal and lack of turbulence ensures smooth concentration gradients. As the eyes of male bathypelagic fish are well developed, it appears that a visual attraction to the luminous lure may play a part in the final docking procedure. Once a male angler has established itself upon its mate, most of its sense organs, including its olfactory lamellae, degenerate and it becomes little more than a bag of sperms.

Postnatal growth and development of the olfactory epithelium is not restricted to fish. In mammals, the sensory membrane is supported by the endo- and ectoturbinal ridges of the ethmoturbinal bone, and also by the cribriform plate of the ethmoid bone (Fig. 1.3a). In the adult mammal there are six clearly distinct endoturbinal spurs of the ethmoturbinal but, in the juvenile, differentiation of the deepest pair is incomplete (Fig. 1.4). As the young mammal grows, the spurs elongate and deepen and their surfaces become somewhat grooved. The increase in the area of olfactory epithelium is accomplished by a generation of more sensory cells and not by a decrease in density such as would be caused by a thinning out. In both adult and juvenile cattle the density of receptor cells is about six million per square centimetre, but the adult has twice the area of newly born youngsters.

The olfactory epithelium of vertebrates has a structure which shows remarkably little variation between classes. It consists of a basal membrane lying on the turbinal bone which supports three main cell types; the receptor cells, ciliated support cells and basal cells (Fig. 1.5a). In addition there are mucus-secreting Bowman's glands. The whole membrane is about 95 μm thick, being a little thicker in juveniles than adults. The receptor cells are bipolar structures which terminate in a swelling known as the olfactory knob. Arising from the knob are a series of filamentous cilia, the actual number of which appears to be species-specific. In cattle, each knob supports 17 cilia (Menco, 1977). In amphibians the cilia are about 200 μm in length, but in mammals the maximum recorded length is 60 μm (Seifert, 1970). The proximal 2 μm or so is considerably

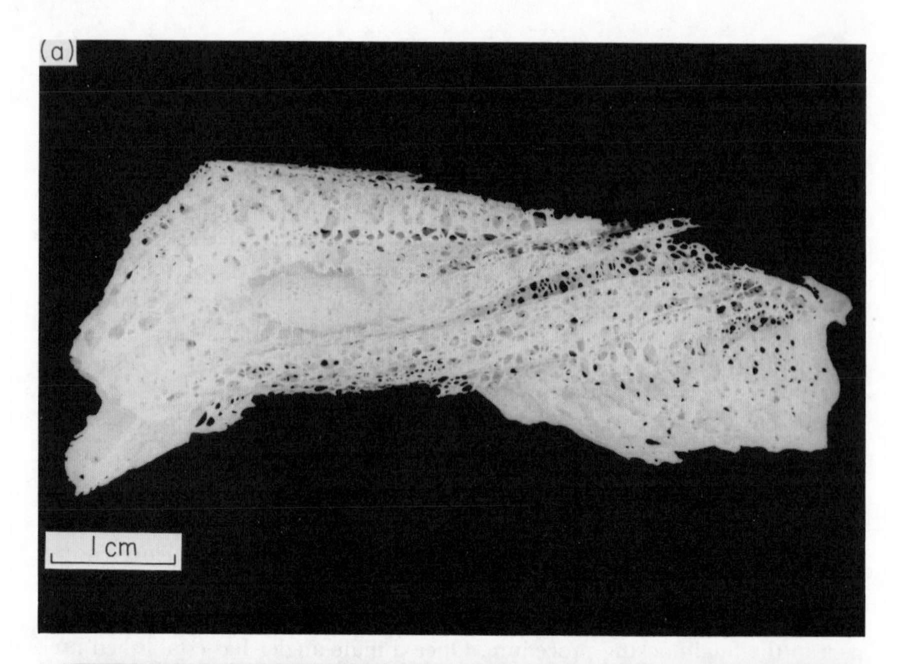

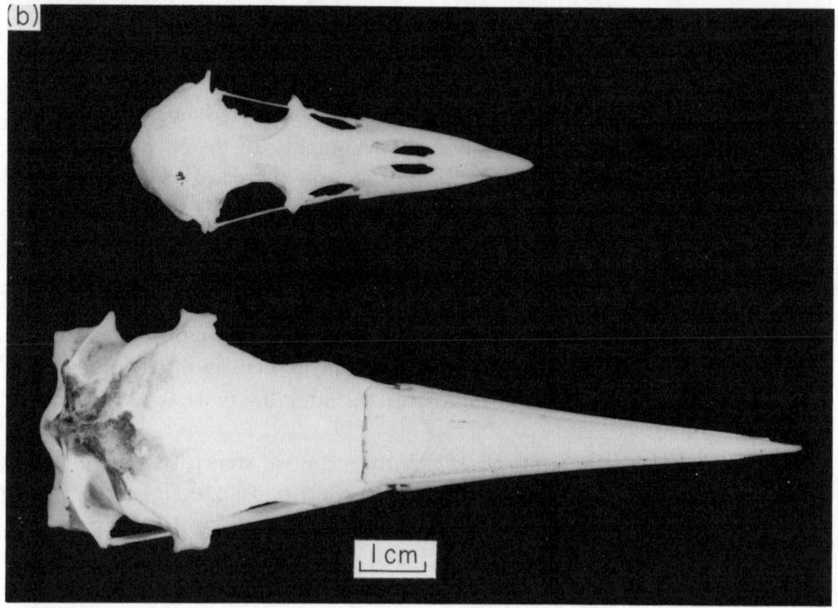

Figure 1.3
(a) Right turbinal bone complex of an adult grey kangaroo *Macropus giganteus*. (Anterior to right.)
(b) Skulls of fulmar petrel *Fulmarus glacialis* (above) and gannet *Sula bassana* (below), from above. The absence of any trace of external nares in the gannet can easily be seen by comparison with the fulmar.

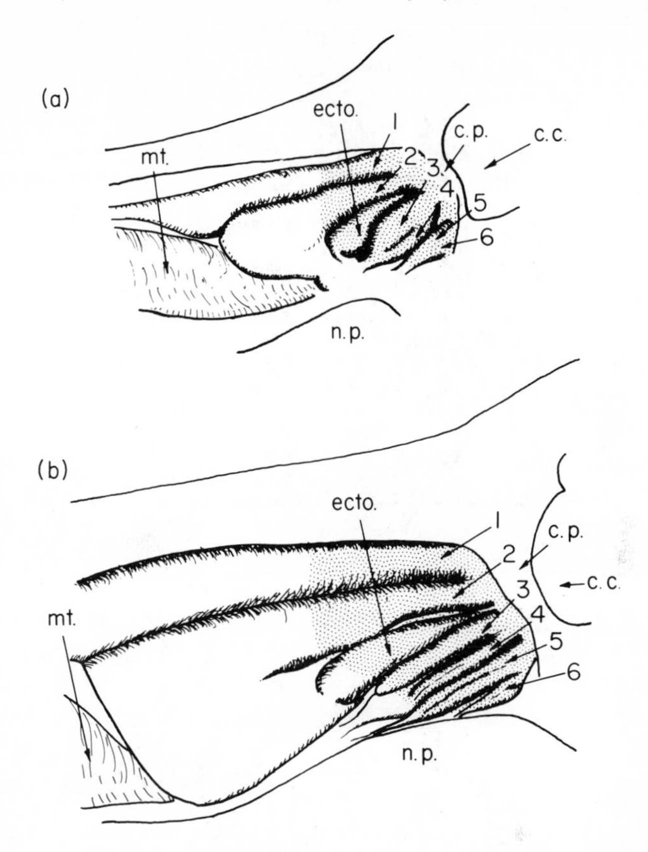

Figure 1.4 Diagram of sagittal section through skull of: (a) one-week-old calf and (b) two-year-old steer, to show the ethmoturbinal and extent of the olfactory epithelium, after removal of the nasal septum. c.c. cerebral cavity; c.p. cribriform plate; ecto. ectoturbinal part of ethmoturbinal; mt. maxilloturbinal; np. nasopharynx; 1–6 divisions of the endoturbinal part of the ethmoturbinal, olfactory epithelium. (After Menco, 1977.)

thicker, with a diameter of 0.2 μm, than the long distal taper, diameter 0.08 μm, and perhaps this plays a support role, for shortly after the taper starts the cilia turn through 90° and spread out over the epithelium in a vermiform sheet (Fig. 1.6). Together with the mucus in which they are bathed, they form a layer about 5 μm thick. Since the cilia are the first parts of the nervous system encountered by an incoming odorant molecule, it would seem reasonable to expect that specific adsorption sites might be present on their peripheral membranes. The technique of electron microscopy called freeze-fracture reveals the existence of particles which are proteinaceous in nature on the membranes (Menco 1977); and since there is physical evidence that adsorption sites must be proteinaceous (Getchell

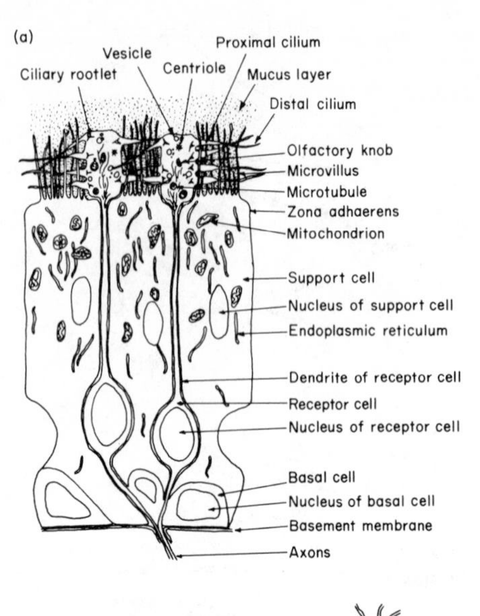
(a)
Vesicle
Proximal cilium
Ciliary rootlet
Centriole
Mucus layer
Distal cilium
Olfactory knob
Microvillus
Microtubule
Zona adhaerens
Mitochondrion
Support cell
Nucleus of support cell
Endoplasmic reticulum
Dendrite of receptor cell
Receptor cell
Nucleus of receptor cell
Basal cell
Nucleus of basal cell
Basement membrane
Axons

(b)

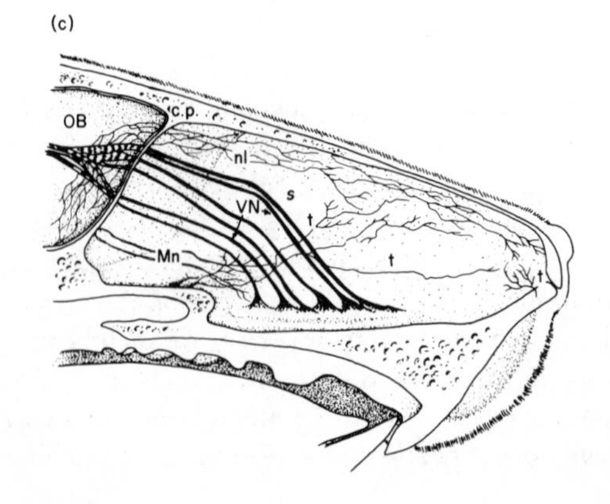
(c)
OB
c.p.
nl
s
VN
t
Mn
t
t

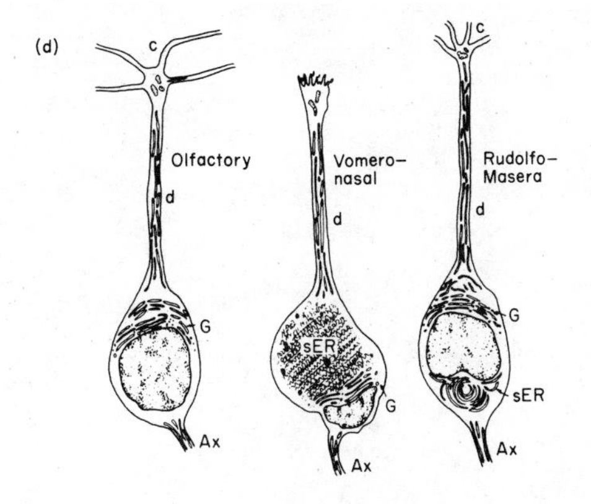

Figure 1.5
(a) Schematic representation of the components of the olfactory epithelium.
(b) Stages in the differentiation, maturation and senescence of an olfactory receptor cell. The sequence starts with a basal cell on the left and ends with a degenerating bipolar cell on the right.
(c) Sagittal section through the nose of a rat. o.b. olfactory bulb; n.t. terminalis nerve; v.n. vomeronasal organ of Jacobson; m.n. nerves from the organ of Rudolfo-Masera; c.p. cribriform plate of the ethmoid bone.
(d) Diagrammatic representation of a receptor cell from the olfactory epithelium, sensory lining of the vomeronasal organ, and from the organ of Rudolfo-Masera. c. cilial bases; d. dendrite; s.e.r. smooth endoplasmic reticulum; ax. axon; g. Golgi apparatus.
(a) from Stoddart, 1976 (a); (b), (c) and (d) from Graziadei, 1977.

and Gesteland 1972), these particles may function in odour reception. After the odorant has triggered the adsorption site, the impulses travel along the cilium towards the knob, but a single triggering may influence more than just one cell. At intervals along the cilium are swelling or vesicles which may coalesce with others from other cilia. Furthermore, the very tip of the taper carries a vesicle which frequently can be seen to be closely appressed to the terminal vesicle of another cilium.

An interesting feature of the olfactory nervous system is that the receptor cells disintegrate after between 28 and 35 days and are replaced by fresh ones which develop from the basal cells of the epithelium (Fig. 1.5b) (Moulton, 1974; Graziadei and Moulton, 1977). Considering the virtual absence of cell renewal in the nervous system generally, the behaviour of the olfactory epithelium is quite remarkable.

Not all the air inspired is subjected to olfactory examination; usually only quite a small proportion actually passes far enough upwards and backwards in the nasal cavity to reach the olfactory cleft (Fig. 2.3). The proportion can be altered by the expansion and contraction, in mammals at least, of the swell bodies, vascular valve-like structures lying at the base of and at either side of the anterior part of the nasal septum. When they are contracted the inspired air

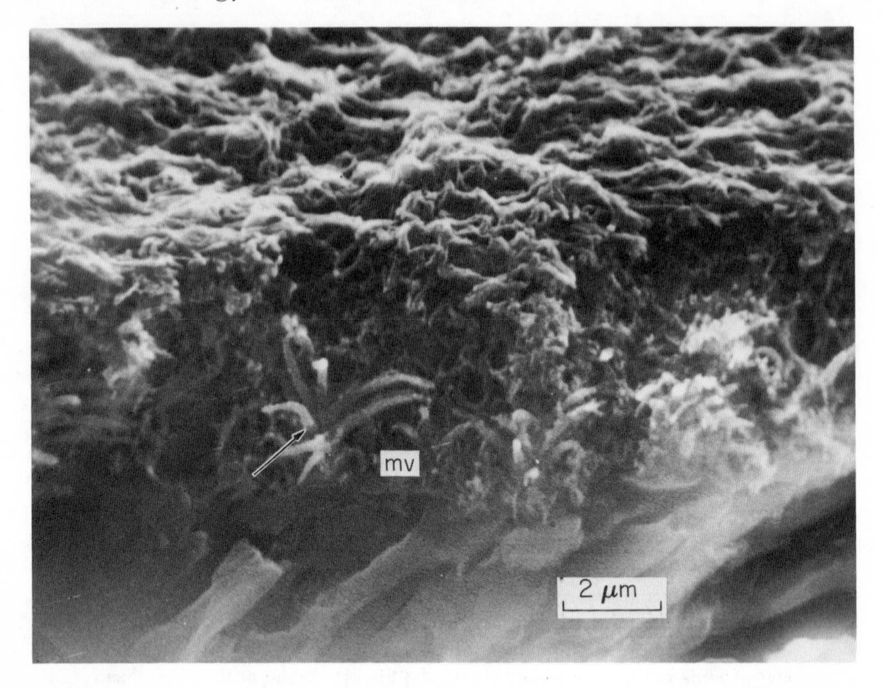

Figure 1.6 Scanning electron micrograph of the edge of a crack in the olfactory epithelium of a calf. Note the dendritic nerve ending (arrowed) and the thick mass of distal cilia overlying the membrane and bathed in mucus. Some supporting cell microvilli (mv) can be seen close to the nerve ending. (Photograph: B. P. M. Menco)

passes directly backwards towards the trachea; when expanded, the air stream is directed upwards through the maxilloturbinal complex, there to be cleaned, heated or cooled and saturated with water so that the odorant molecules may be deposited on the olfactory mucosa.

The anatomical structure of the olfactory chamber of a series of representative vertebrates is shown in Fig. 1.7. In all terrestrial species the chamber lies in the respiratory passage, but in fishes it occupies a position in front of the eyes and has no contact with the respiratory system. Such an arrangement has certain advantages which will be discussed later.

Associated with the vertebrate olfactory epithelium is a pigment which is pathologically black in fishes, amphibia and reptiles, and brown to brownish-yellow in mammals. In fish and amphibia there are distinct pigment cells lying in the epithelium, but in mammals the pigment seems to be associated with the support cells and Bowman's glands. For many years, it was thought that albino mammals had an impaired sense of smell, so explaining the deaths of albino domestic stock from their failure to identify poisonous plants (Negus, 1958). Their deaths must have been caused by some other deficit related to albinism, for

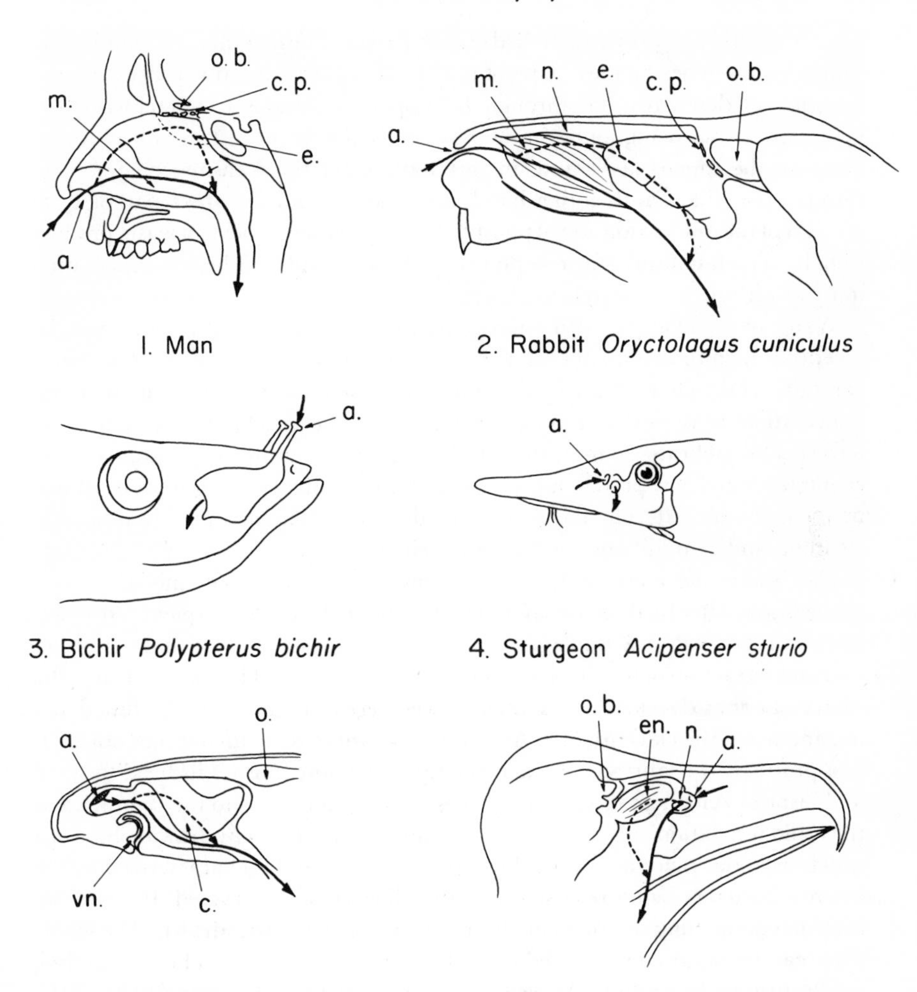

Figure 1.7 Diagram of the passage of air through the nasal structures of various vertebrates. Solid arrow denotes passage of respiratory stream; broken line denotes passage of fraction directed over olfactory epithelium. a. external naris; o.b. olfactory bulb; e. ethmoturbinal; en. endoturbinal; n. nasoturbinal; m. maxilloturbinal; vn. vomeronasal organ; c. concha; c.p. cribriform plate. (After Allison, 1953; Negus, 1958; Kleerekoper, 1969.)

carefully controlled laboratory studies indicate that there is no difference in olfactory acuity between albino and pigmented rats. Furthermore, it appears that the olfactory pigment is controlled by different alleles from those controlling albinism, for albino rats and rabbits have normally pigmented olfactory epithelium (Jackson, 1960; Moulton, 1960; Ottoson, 1963). The function of the pigment is the subject of a great deal of speculation. The fact that it is found in the support cells suggests it may help in the olfactory nerve excitation process, for the receptor cell dendrite passes through the support cells and is in contact with their cytoplasm. Also, being found in the Bowman's glands suggests that it may also occur in the mucus layer and, in this way, affect the transmission processes through the cilia. On the other hand, the yellow colour may be a pathological artefact of the auto-oxidation of phospholipids, themselves being a waste product from lipid metabolism of the receptor cells (Jackson, 1960). The real explanation must await further researches.

Water or air charged with odorous particles does not simply wash over the receptive membranes in the olfactory organ. It is injected and guided in a carefully controlled manner, the amount sensed increasing as more exact information is required. Enough is known of the mechanics of sniffing in terrestrial vertebrates to indicate that there is no universally applicable series of events, although the process is broadly the same in all forms. Air must be drawn in through the external nares, presented to the olfactory membranes after cleaning and humidifying, and then expelled.

The ability to increase greatly the inspiration rate, polypnoea, is seen immediately after birth in the rat, but three other behavioural sequences develop somewhat later. Protraction and retraction of the mystacinal vibrissae, protraction and retraction of the tip of the snout, and orientation of the head towards the source of odour develop during the first week of life and are all fully functional on, and from, the eleventh day. As a young rat approaches an odorous object, it first protracts the vibrissae, retracts the tip of the nose, pushes its head forward and inspires. The nose retraction exposes the flaring nostrils to the stimulus and the rapid inhalation aids olfactory organ stimulation. The withdrawal phase of a sniff lasts about twice as long as the approach phase and is characterized by the above characters being reversed, i.e., the vibrissae are retracted, the nose tip protracted, the lungs emptied of air and the whole head withdrawn. The whole cycle can be rapid and repeated up to 11 times a second, and a bout of sniffing can last up to 10 seconds. Movements of the vibrissae are not essential to the process for, in a low-intensity sniff, only the nose tip movement and polypnoea may be seen; the olfactory function of vibrissae is not clear. Sniffing in many species of mammals occurs without the use of vibrissae; for example, in the guinea pig, *Cavia*, capybara, *Hydrochoerus*, ground squirrel, *Geomys*; cat, *Felis*; dog, *Canis*; raccoon, *Procyon*; coati, *Nasua*; panda, *Ailurus*; and in the fur seal, *Callorhinus*, and common seal, *Phoca*. The rat-like sniff is reported to occur in the mouse, *Mus*; gerbil, *Meriones*; chinchilla, *Chinchilla*; flying squirrel, *Glaucomys*; and the golden hamster, *Mesocricetus* (Welker, 1964).

Although some parts of the sniffing cycle are seen from birth, young rats do not show the entire cycle until their third week of life. By then about 4.0% of their daily activity is active sniffing. Most of the rest of the day is spent feeding or sleeping (Bolles and Woods, 1964). Fishes, amphibia, birds and reptiles have simple external nares lacking any structural mechanism allowing such a sniff cycle. Amphibia and aquatic mammals close the nostrils prior to diving – a phenomenon seen in partially aquatic reptiles, but not seen in aquatic birds. There is no evidence of any special sequence of events correlated with odour perception in reptiles, with the exception of a projection of the head and a slight increase in inhalation rate.

By far the majority of vertebrates have paired external nares through which air, or water, is drawn for odour detection. The exceptions to this are the lampreys and hagfish, which have a single naris formed from fusion of the basic pair, and several species of birds which lack external nares altogether. These birds belong to the families Sulidae and Phalacrocoracidae which are composed of high-diving, fish-eating marine seabirds (gannets, boobies and their relatives, and cormorants, shags and snake-birds, respectively). To obviate the problem of inhalation caused by the disappearance of the nostrils, secondary nares have developed at the angle of the mouth, permitting transoral breathing (Fig. 1.3b and 1.8f). Open nares on the bill are clearly disadvantageous to high-diving birds and a partially open mouth only a little less so. By manipulation of the jugal bone and its associated horny operculum, the secondary nares are able to be shut immediately prior to diving (Macdonald, 1960). The inspired air is unable to reach the olfactory vesicle so, in these birds, the whole system is rendered obsolete, even though their olfactory bulbs, the anterior lobes of the brain which process stimuli from the olfactory receptors, are quite well-developed (Bang and Cobb, 1968). A stage in the evolution of this characteristic is seen in the Pelicanidae and, to a lesser extent, in the Spheniscidae. The horny rhamphotheca, or bill covering, of pelicans has only a tiny narrow slit over the nares. It is doubtful if this has more than rudimentary function. A small flap of skin partially occludes even this meagre opening, and when the bird dives water pressure causes the flap to seal the slit. A similar situation exists in penguins, but the slit, and its valve-like flap, are somewhat larger.

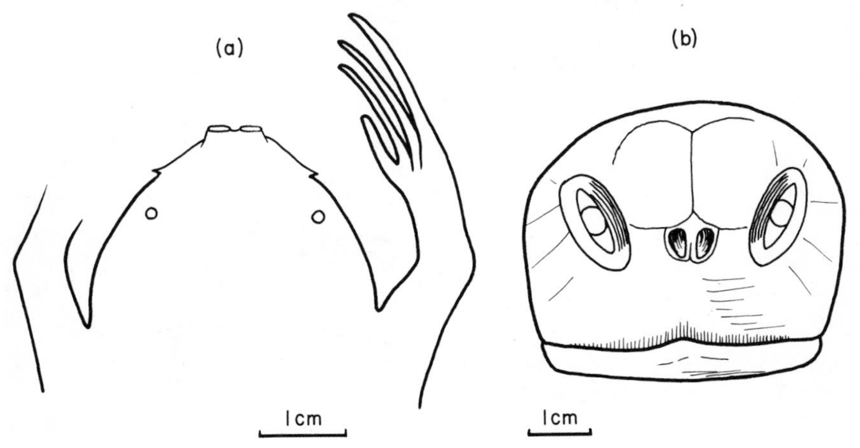

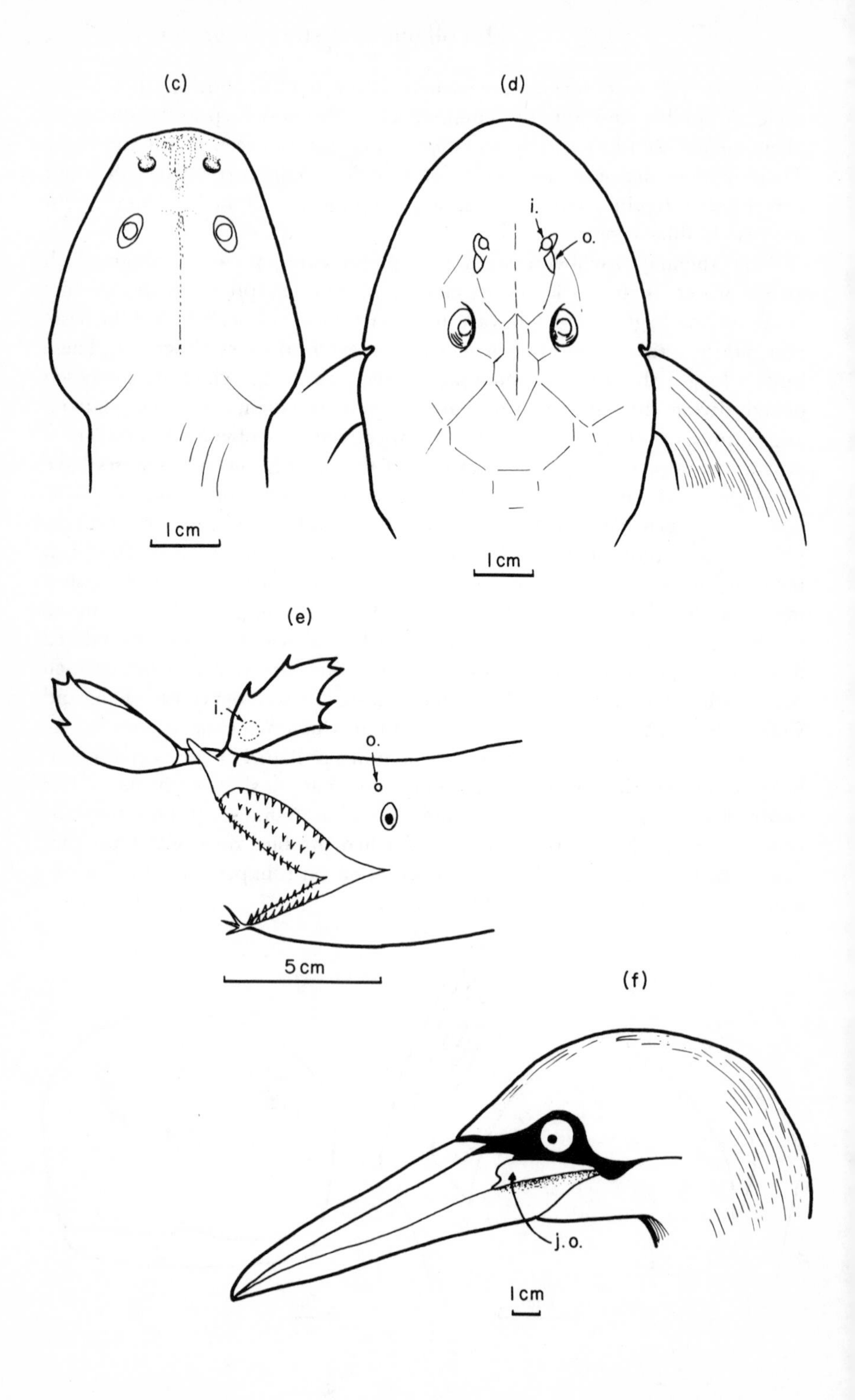

(c)
(d)
i.
o.
1 cm
1 cm
(e)
i.
o.
5 cm
(f)
j. o.
1 cm

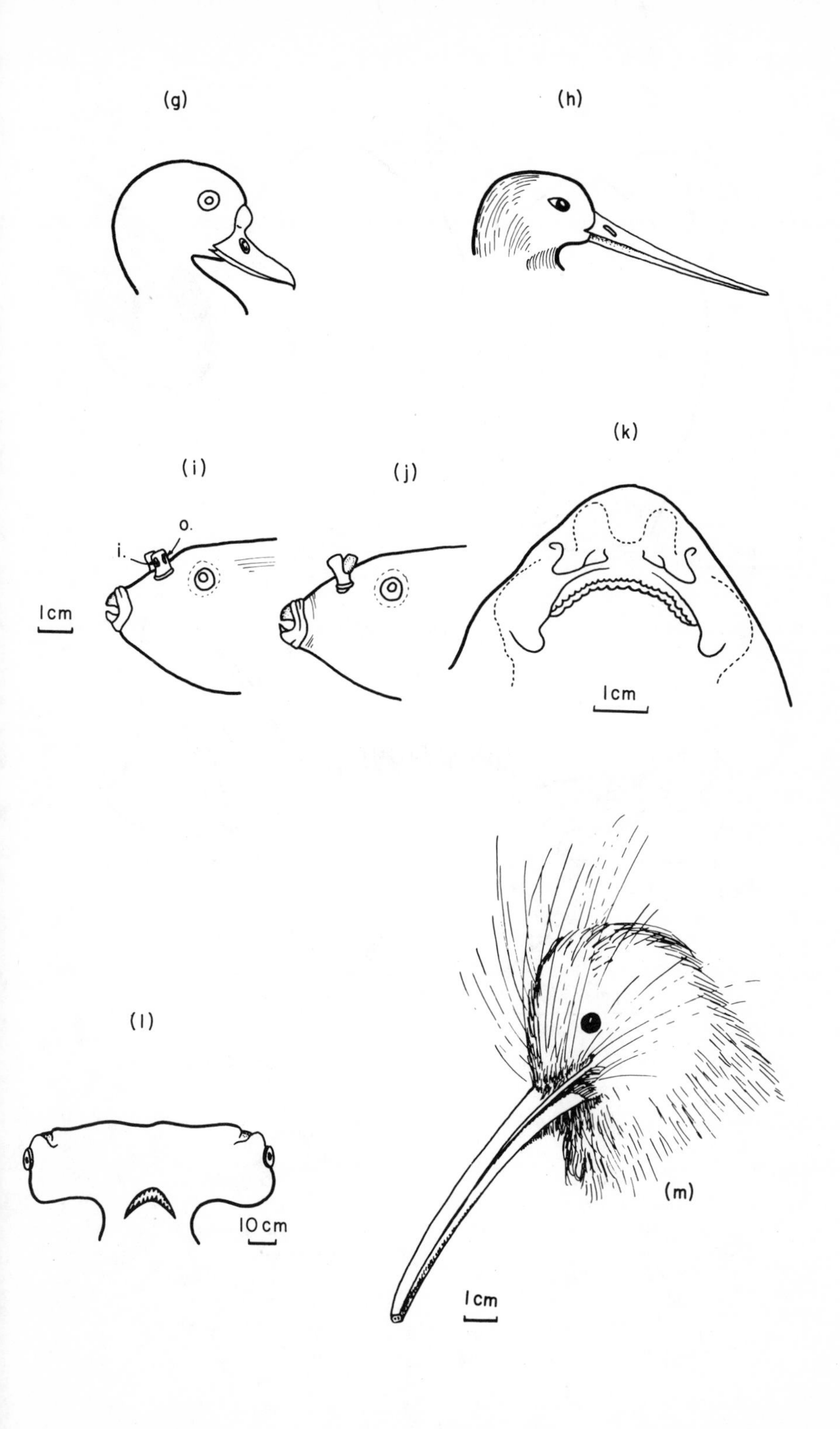

(g)
(h)
(i)
(j)
(k)
o.
i.
1cm
1cm
(l)
10cm
(m)
1cm

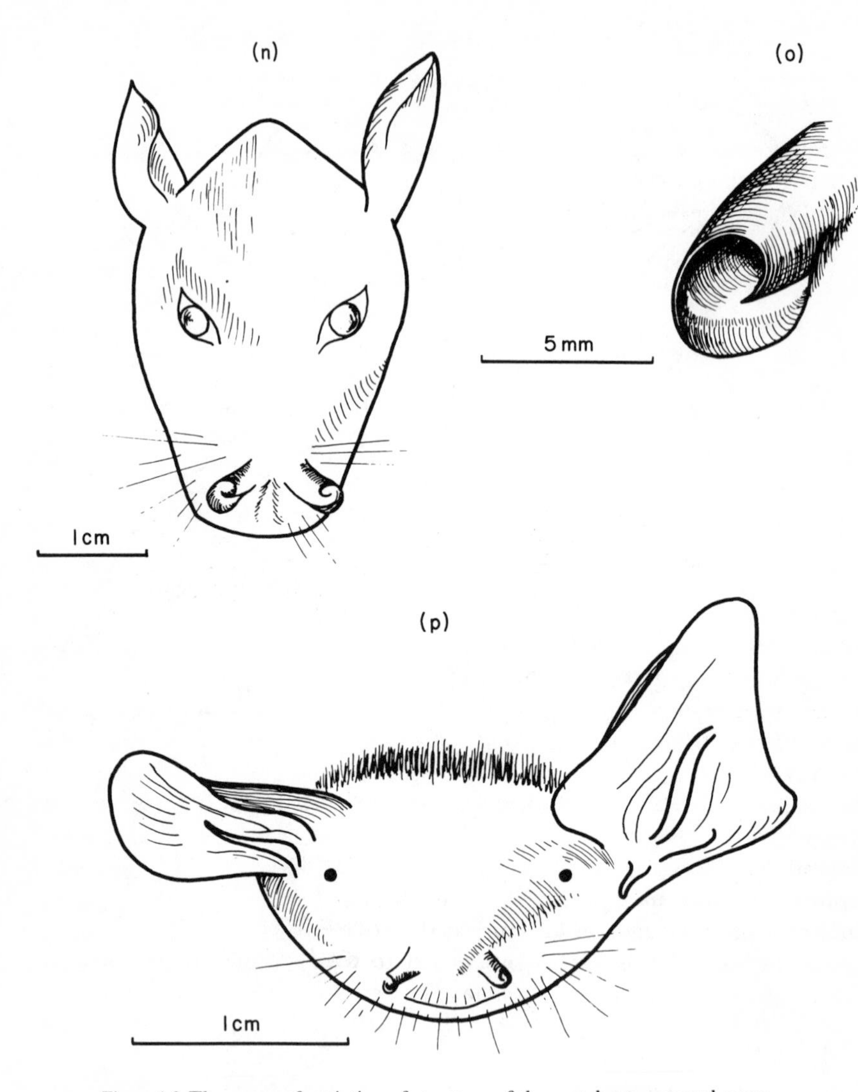

Figure 1.8 The range of variation of structure of the vertebrate external nares.

(a) Surinam toad *Pipa pipa* (1)

(b) Tortoise *Chelone* sp. (2)

(c) Death adder *Acanthophis antarcticus* (3)

(d) Mailed catfish *Plecostomus plecostomus* (1)

(e) Moray eel *Rhinomuraena* sp. (6)

(f) Gannet *Sula bassana* (1)

(g) Shelduck *Tadorna tadorna* (3)

(h) Godwit *Limosa lapponica* (3)

(i) Parrot fish *Tetrodon pardalis* (2)

(j) Parrot fish *Tetrodon nigropunctatus* (2)

(k) Dog fish
Scyliorhinus caniculus (1)
(l) Hammerhead shark
Sphyrna tudes (4)
(m) Kiwi
Apteryx australis (1)
(n) Tube-nosed fruit bat
Nyctimene aello (5)
Specimen number BM (NH) 73.2031
(o) Tube-nosed fruit bat
Nyctimene aello (5)
right narial tube.
Specimen as in (n)
(p) Tube-nosed insectivorous bat
Murina huttoni (5)
specimen number BM (NH) 67.1606

(1) drawn from museum specimen, Zoology Department, King's College, London
(2) drawn from museum specimen, Zoology Department, Monash University, Melbourne
(3) drawn from life
(4) from Bigelow and Schroeder, 1949
(5) drawn from museum specimen, British Museum (Natural History)
(6) after Grassé, 1958
i. inlet; o. outlet; j.o. jugal operculum.

The variation of form of the external nares of vertebrates is very great and it is seldom clear what factors have provided the selective pressures moulding the final result. In some species the nares have been much affected by changes wrought by non-olfactory evolutionary developments such as those seen in elephants, anteaters, echolocating microchiropteran bats, and aquatic species for which a nostril-closing device is necessary. In others, they assume a variety of forms which are not obviously associated with any particular structural developments to the front of the head. Many fish, for example, have tentacle-like narial structures (e.g., *Tetrodon*, *Rhinomuraena*); others have complex hooked flaps, called pavilions, separating inhalent and exhalent pores. Most amphibia and reptiles have plain nares set flush on the front or top of the head, but few mammals have nares as simple in structure as those of man or pig. In this class of vertebrates the narial walls are usually curled to some extent and in non-echolocating megachiropteran bats they may be extended into long, curled tubes (Fig. 1.8, 1.15). Some of these adaptations may be related to possible stereolfaction (see Section 1.3).

Before considering the structure of the neural pathway leading to the brain, it is necessary to examine first the accessory olfactory organs. These are structures which are associated with the main olfactory system, but have a separate and distinct innervation. Of these, the most obvious is the organ of Jacobson, sometimes called the vomeronasal organ (Fig. 1.9). In elasmobranchs and the Dipnoi the organ is only partially separated from the main olfactory cavity, but it has its own nerve, the vomeronasal nerve, which passes to the accessory olfactory bulb of the brain. The separation of the organ from the olfactory cavity is more complete in the teleost fish and urodele amphibians and can be first considered to be quite complete in the anuran amphibians. In the reptiles (Fig. 1.7), the organ

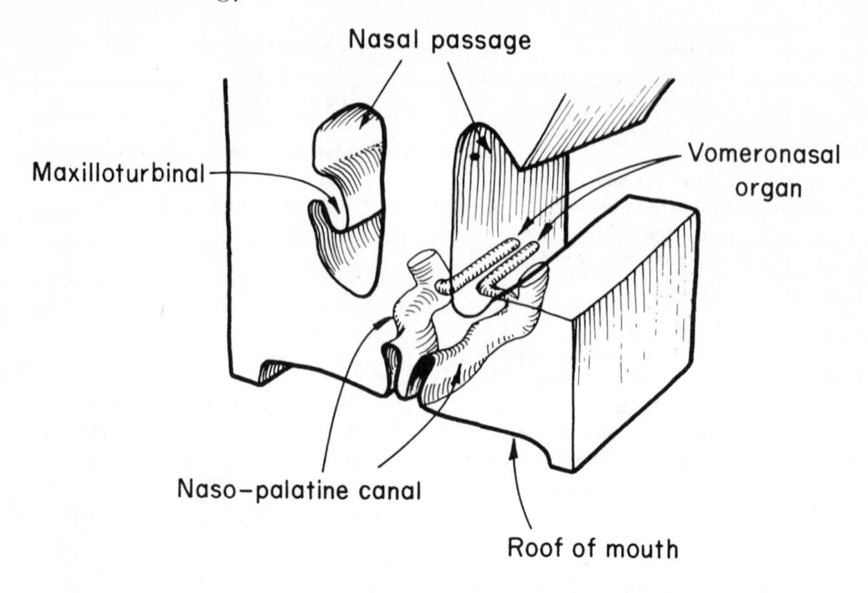

Figure 1.9 Block diagram of the vomeronasal organ, nasal passage and naso-palatine canal of the hedgehog, *Erinaceus europaeus*. (After Poduschka and Firbas, 1967.)

is well-developed (though it is absent from chameleons), having severed its connection with the olfactory cavity in favour of a duct through the hard palate into the roof of the buccal cavity. In most species it is a paired blind diverticulum but in the higher reptiles, the crocodilia, the lachrymal duct opens into its posterior end. The organ is absent in birds but is well-developed in mammals (Stephan, 1965). It lies at the junction of the nasal septum and vomer bone, embedded in a bony or cartilaginous capsule, and opens to the roof of the mouth via the naso-palatine canal of Stenson. Part of the vomeronasal organ is lined with olfactory epithelium, but the medial aspect bears highly vascularized tissue. It is thought that inflation and deflation of these blood vessel complexes brings about a pumping motion capable of filling and emptying the organ with odour-laden air or water (Fig. 4.8) (Allison, 1953). The receptors of the vomeronasal organ respond to the same odorants as the olfactory mucosa proper, although at different thresholds (Tucker, 1963; Moulton and Tucker, 1964). Graziadei indicates that in the turtle, *Terrapene carolina*, a severed piece of the vomeronasal nerve was soon bridged by regenerating axons. Further studies with mice reveal that receptor cell renewal occurs much as it does in the nose (Graziadei, 1977).

Recent anatomical studies using electron-microscopy techniques have indicated the existence of three other accessory olfactory organs in mammals. These are the septal organ of Rudolfo-Masera; the terminal endings of the *nervus terminalis*; and the terminal endings of the *nervus trigeminus* (Vth cranial nerve). The relationship of all five neural components of the olfactory system of a rat is shown in Fig. 1.5c. The organ of Rudolfo-Masera consists of a small patch on the ethmoturbinales which gives rise to two discrete nerves which do not fuse or interchange fibres with the olfactory or the vomeronasal nerve bundles. The receptors broadly resemble those found in the true olfactory region; the main differences affect the arrangement of organelles (Fig. 1.5d). The *nervus terminalis* is a complex of fine nerve bundles passing from the nasal septum to the olfactory bulb. Like the nerves supplying the organ of Rudolfo-Masera, they maintain their independence throughout their distribution. Finally, the distal portion of the *nervus trigeminus*, which perfuses much of the anterior portion of the nasal cavity, is known to respond to odorant molecules. Other portions of the Vth cranial nerve serve motor and other sensory functions; unfortunately nothing is known of the specific capabilities and role of these little-understood accessory units of the olfactory system (Tucker, 1963; Bojsen-Møller, 1975; Graziadei, 1977).

The olfactory bulbs in primitive vertebrates lie anterior to the brain, connected to it via the olfactory nerve (Fig.1.10). In higher vertebrates, they come to lie underneath the brain, overlain by the cerebral hemispheres. The olfactory bulbs function as a special 'odour brain' to process the signals before they reach the brain proper (Pfaffman, 1971). The relative size and position of the olfactory bulbs of several species of vertebrates are shown in Fig. 1.11. The axons of the receptor cells terminate in the glomeruli, discrete clusters of neuroterminals, which, when viewed all together, constitute the *stratum glomerulare*. In the rabbit, one of the very few species in which the total numbers of the various components of the olfactory system have been calculated, 100 million primary receptors feed into 3800 glomeruli (Allison and Warwick, 1949). From each glomerulus emerge 24 output cells (unmyelinated), which soon take a myelin sheath and are known as mitral cells, and a bank of 24 tufted cells which pass to the other olfactory lobe (Hainer, Emslie and Jacobson, 1954). The convergence ratio between receptor cells and secondary olfactory neurons is very high; in some species of bats it is 900 : 1 and in the burbot, *Lota lota*, it is 1000 : 1 (Bhatnagar and Kallen, 1975; Gemne and Døving, 1969). What is the advantage of such numerous receptors? Van Drongelen *et al.* (1978) have shown theoretically that convergence may be responsible for the high sensitivities found in the olfactory system. A single mitral cell might become maximally depolarized at quite low stimulus levels due to the spatial summation of many receptor cells. The ultimate limiting factor to the system sensitivity is the absolute number of odour acceptance sites and the probability of response of the receptor neurons. It is known that the receptor cells of the frog are multimodal; that is, they respond to more than just a single odorant type. Theoretically, such cells have a smaller response probability than unimodal ones, but if the number of sites is high enough there will be an adequate compensation.

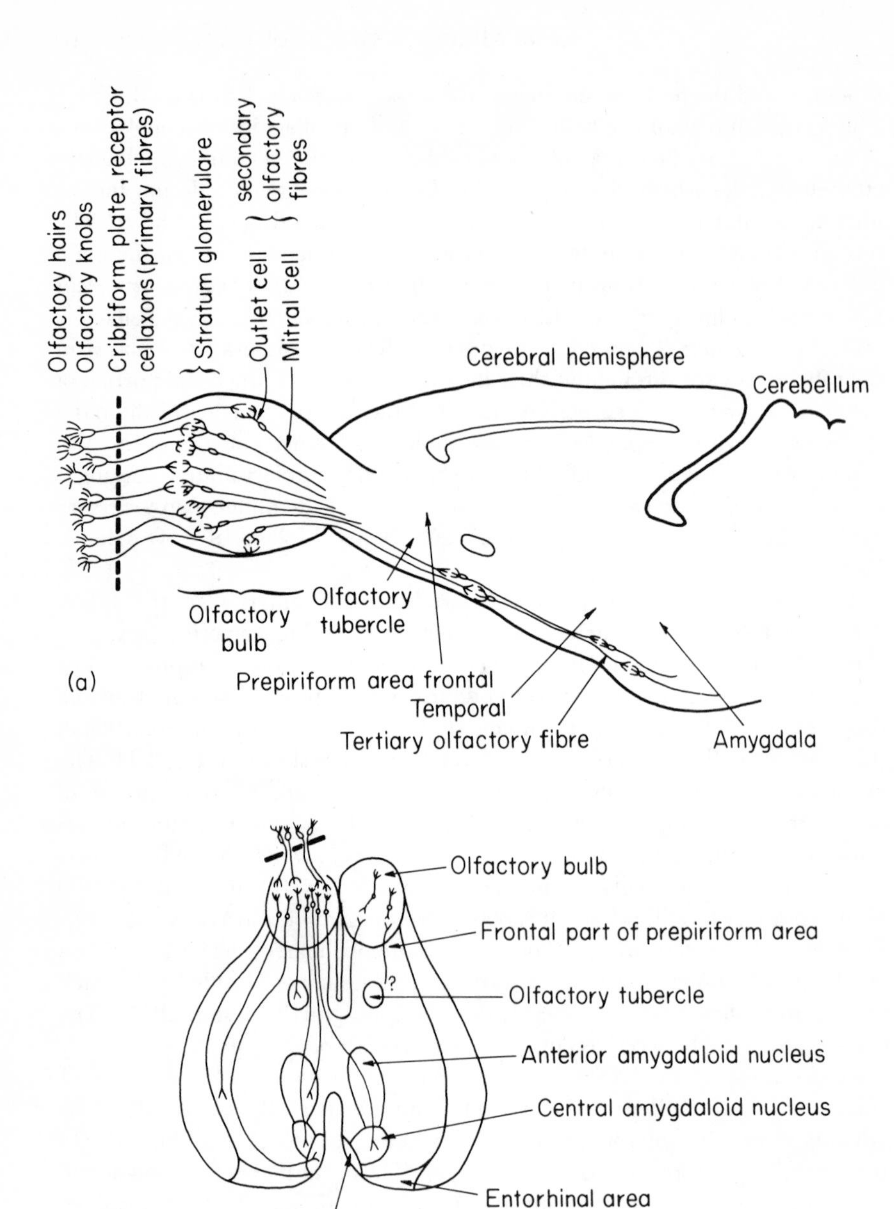

Figure 1.10

(a) Diagrammatic LS representation of the neural pathways in the olfactory system of the rat. (After Nieuwenhuys, 1967; Allison, 1953; Hainer, Emslie and Jacobson, 1954.)

(b) Diagrammatic TS representation of the secondary olfactory pathways in the mammalian brain. (From Allison, 1953.)

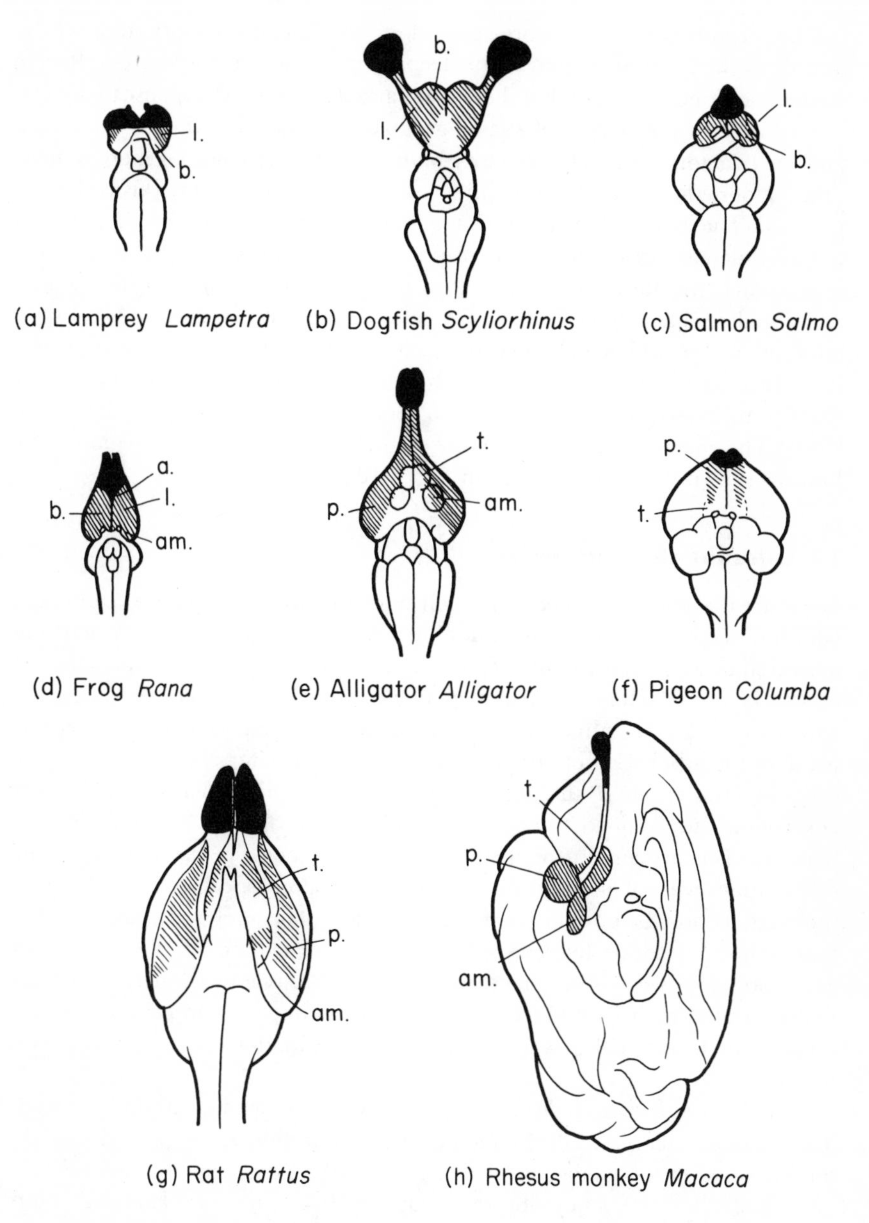

Figure 1.11 Ventral view of the brains of a sample of vertebrate species showing the extent of the olfactory bulb (solid black) and secondary olfactory fibres (hatched). (h) is shown in sagittal section.

l. lateral olfactory area; b. basal olfactory area; a. accessory olfactory bulb; am. amygdala; p. prepiriform area; t. olfactory tubercle. (Redrawn from Allison, 1953.)

The secondary olfactory fibres pass along the lateral olfactory area on the outside of the pre-optic region to the deep layers of the olfactory system. Recent studies have demonstrated that there is a marked differential response to odours, as measured by the activity of a small group of cells, between the pre-optic area and the olfactory bulb. The olfactory bulb of the rat responds far more to non-urine odours than does the pre-optic area, but for urine odours this is sharply reversed. The hypothalamus, which together with the prepiriform complex and the amygdaloid nucleus forms the pre-optic area, is intimately involved in the regulation of the hormones produced by the anterior part of the pituitary gland. As we shall see in Chapter 4, urine of females of many vertebrate species contains odorous factors which influence the sexual behaviour of males and govern courtship, and the pre-optic area is known to control sexual behaviour (McCann, Dhariwal and Porter, 1968; Scott and Pfaffman, 1967; Pfaffman, 1971). The olfactory pathway from the epithelium to the pre-optic region of the brain is shown diagrammatically in Fig. 1.10.

1.2 Odorant characteristics

What are the properties of odorant molecules and how do they differ from non-odorant ones? This intriguing question seems likely to remain largely unanswered until the cause of the depolarization of the end organs in the olfactory organs in the olfactory epithelium is finally elucidated. There is widespread agreement that the initial event in the process leading to odour perception involves the adsorption of odorant particles onto specific sites on the receptor cells, but this is where the agreement ends. It is quite unclear whether the adsorption sites respond to chemical or physical properties of the molecules. Some workers argue that the membrane of the receptor cell is permeable and differential levels of permeability bring about odour specificity. It has been shown that odorous substances act as accelerators of haemolysis by saponins and that a linear relationship exists between the logarithms of the haemolytic accelerating powers and olfactory thresholds (Davies and Taylor, 1958). Different thresholds for different odorants depend upon the adsorption energies involved in the molecule passing from the air to the lipid-aqueous interface of the membranes.

A characteristic of all molecules is that they vibrate with a certain frequency. The original champions of the molecular vibration theory suggested that the quality of odour of a substance was determined by its 'osmic frequency' – the pattern of its low frequency vibrations (Dyson, 1938). Other workers have, more recently, criticized this theory because the range of frequencies considered could only be related to the presence in the molecule of certain chemical groups. Since theories based on the presence of certain chemical groups have been largely discredited, this vibration theory has been discredited too. In its place these workers suggest that the low frequency vibration of the whole molecule may be the specific factor. They have amassed much corroborative data, but there is still much speculation (Wright and Burgess, 1975; Wright, 1976, 1977).

Against this very uncertain backdrop of ideas, it is difficult to point to the precise properties of odorant molecules. There is general agreement that to be odorous a substance must be volatile, and it must be at least partly soluble in both water and lipid. That physical, rather than chemical, properties of a molecule are responsible is suggested by experiments upon olfactory potency of members of homologous series (Passy, 1892; Ottoson, 1958). As the series ascend, whether it be primary aliphatic alcohols, aldehydes or fatty acids, the threshold at which the substance is perceived decreases. This has been shown for a wide range of animals, both invertebrate and vertebrate (Ottoson, 1963). The increase of odorant potency in these series might be related to any of the physical properties of vapour pressure, water and lipid solubility. Ottoson argues that the relative saturation of the vapour of a substance in the vapour phase represents its thermodynamic activity and, since this is the same in all phases in an equilibrium system, the activity at the point where the biological action is exerted can be determined by measuring the activity in the vapour phase. If the action of an odorous compound on the receptor surface represents an equilibrium system, the potency of different substances might be analysed in terms of thermodynamic activity. He showed that alcohols of intermediate length and equal thermodynamic activity were equally potent, though some members of the series do not fit this trend. Clearly then, some factors other than thermodynamic activity must influence the end organ discharge (Ottoson, 1958).

1.3 Function of the external nares and related structures

Odours convey a great many sorts of information to animals and, in many situations, the quality of the information is significantly enhanced if the source of its emanation can be quickly pinpointed. Vertebrates employ a variety of mechanisms to effect orientation to an odour stream, the chief amongst them being:

1 The enhancement of perceptual asymmetry in the paired olfactory chambers, either by differentially modifying the flow rate through each chamber, or by increasing the time-lag between the odour striking first one, then the other, nostril.
2 Sampling of a broad front by swinging the head from side to side – klinotaxis.
3 Keeping the head steady but sampling air from two points as widely separate as possible – tropotaxis. This necessitates specialized sampling structures. The nose certainly does not function as a single entity, and some species utilize more than just one of the mechanisms outlined above to introduce a directional component into their perception.

Many, and probably most, species of mammals possess vascular swell bodies lying on the maxilloturbinal bones in the anterior part of the nasal passage. Their cyclic distension and relaxation creates and maintains substantial differences in the amount of air sampled by each passage (Fig. 1.12) (Bojsen-Møller and Fahrenkrug, 1971). By enhancing the differences in receptor input

Figure 1.12 Series of expirations made by an anaesthetized rat on a moving cold metal mirror (Zwaardemaker mirror). Note the greater expiration flow from the right side of the nose. (From Bojsen-Møller and Fahrenkrug, 1971). Reproduced by kind permission of Cambridge University Press.

sensation, this cyclical phenomenon provides a means whereby the animal is able to isolate, at a perceptual level, the invariants (such as the odour quality) of the input, without having to move the receptors about very much. The asymmetry of the stimulation enables accurate comparisons to be made between the right and left, which are necessary if the animal is to orientate towards an odour stream borne on the wind, diffusing outwards from a point source or forced into an upwelling by a sudden, sharp sniff. Cyclic distension and contraction of nasal swell bodies has only been observed in man, rat and rabbit, but since all mammals possess a vascular lining to the nasal bones, all might be subject to periodic changes in nasal resistance. As Bojsen-Møller and Fahrenkrug (1971) put it:

> Functionally, the nose must therefore be considered to consist, at least at certain periods, of two alternating rather than parallel air passages. This should be borne in mind . . . in investigations of . . . olfactory function. Any data referring to nasal function must therefore take into account its periodicity.

The nose must also be considered as having two quite separate points of entry and its stereolfactic ability will be enhanced if the distance between them is great, or if each draws air from the side rather than the front of the animal.

It is unusual for an animal ever to be in still air, with the exception of the time when it is asleep within the safety of its nest or burrow. Most odours arrive at the animal borne on the wind and the direction of the site of their production can be deduced from the time-lag between their arrival at first one, then the other, nostril. Nasal architecture is clearly important here, but no attempts have yet been made to quantify the effect of certain physical structures on odour stream localization. Laboratory experiments with humans indicate that quite short latencies cause perceptual shifts in the apparent direction from which an odour stream flows. In these experiments, air delivery tubes were inserted into the subject's nostrils and the time-delay in sending a burst of odorant into each tube controlled by a pulse switch. A delay of as little as 0.1 mS could clearly be detected, and a delay of 0.3 mS resulted in the sensation switching from one side of the nose to the other. An odour cloud must be visualized as having a leading edge conforming to the laws of aerodynamics. There will be a certain amount of mixing and hence dilution of the odorant, and eddies and vortices will be created according to the prevailing climatic conditions. As far as the perceiving individual is concerned, the leading edge of a cloud is characterized by a rapid build-up in odorant concentration. If humans are presented with odours into both nostrils simultaneously, but the concentrations differ by as little as 10%, the odour will appear to have arisen from the side detecting the higher concentration (von Békésey, 1964).

For the tetrapods, which have the olfactory system lying in parallel, if not actually in series, with the respiratory system, the rate at which air can be sampled is under the control of the breathing reflexes. Fishes have the olfactory apparatus lying in a discrete pouch not in any contact with the respiratory system but, in several families, water is drawn into the pouch by a pumping action of muscles associated with the gill movements, in a manner that quite closely resembles the odour sampling mechanism in tetrapods. In these families, e.g., Percidae, Labridae, Cichlidae, Scorpaenidae, the rate of inundation of the olfactory lamellae is quite high enough to allow any latency in arrival of odorants to be perceived. Both in fishes with this type of odour sampling (collectively called the 'cyclosmates'), and in those in which cilial beating on the lamellae drives water through the pouch (the 'isosmates') the rate of current flow or the speed at which the fish is swimming does not substantially increase the flow over the receptors. In isosmate families, e.g., Salmonidae, Cyprinidae, Siluridae, Gadidae, and Anguillidae, water flows constantly over the lamellae at a rather slow speed; in the tench, *Tinca tinca*, a coloured dye introduced at the inlet to the olfactory sac takes between three and six seconds to be expelled at the exit. This rate of flow is too slow to allow all but the grossest latencies in time of arrival of odorants to be perceived (Døving *et al.*, 1977). Although isosmates may seem to be less well-equipped to use stereolfaction than the intermittently, and fast-sampling cyclosmate species, detailed examination of other aspects of their biology suggests that this is compensated for in several other ways. Many salmonids 'point' towards the source of an odour, maintaining their position in the water by small correcting movements of their fins. The angle of attack of the pectoral fins dictates the speed at which water flows up over the fish's head and into the vicinity of the olfactory sacs (Bardach and Todd, 1970; Bardach and Villars, 1974). Differential movement of the paired pectorals effects subtle changes in the speed at which water is delivered to the left and right receptors and, while this has little effect on the amount of water sampled, it does result in a comparative sampling of two streams of water arriving at the fish with unequal velocities. Any concentration differences in the water surrounding the fish will thereby be enhanced. In the fast-swimming, olfactorily active species such as the skipjack tuna, *Katsuwonus pelamis*, several mechanisms work in concert. A pair of grooves on the front of the fish's head channel water to the incurrent naris; cilial action induces a water swirl around the lamellae, and a hydraulic pump helps accelerate the flow of water (Gooding, 1963).

Some fishes, e.g., the eel, *Anguilla anguilla*; catfish, *Nemacheilus barbatula*; and the bichir, *Polypterus bichir*, have their incurrent nares elongated laterally and forward into tubes. This is an adaptation seen in many vertebrates which serves to widen the base from which odours are drawn, making triangulation of the point of their emanation more precise. In the majority of fishes the tubes are quite plain in structure, but in *Rhinomuraena*, a type of Moray eel, the tubes end in bizarre leaf-like structures which are capable of being turned outwards or upwards, presumably to gain more directional information about an odour source without the necessity for the fish to turn its head (Fig. 1.8e) (Grassé,

1958). The internarial separation for any species lacking tubular nostrils is set by the maximum width of the head which is itself set by the ecological niche the species occupies and its hydro- or aerodynamic requirements. Consideration of the *relative* amount of the maximum head width utilized by the nostrils reveals some interesting trends (Fig. 1.13a). A relationship is apparent in which the nostrils of lower vertebrates are relatively more widely spaced than those of higher vertebrates. This broad difference can be attributed to the increasing flexibility of the cervical region of the vertebral column found in higher forms. The 'stiff-necked' elasmobranchs, actinopterygians and anurans are potentially capable of stereolfaction by virtue of their widely spaced nostrils, while the 'flexible-necked' mammals, birds and reptiles can sample a broad front of air by swinging the head around. This interpretation is complicated by the *actual* differences in internarial separations seen in the vertebrates, for the greater the actual separation the more precisely can the source be pinpointed, unless the odour plume is extremely narrow and localized. Obviously an analysis of relative separation obscures actual differences to be expected in animals of different sizes. Fig. 1.13b analyses the actual separations for small vertebrates (less than 200 grams adult weight) and for the larger snakes. It is at once apparent that fishes, the most stiff-necked of all the vertebrates, have by far the most widely separated nostrils. The great flexibility of the neck of the higher vertebrates means that the external nares can be positioned quite close together and the direction from which an odour emanates estimated by klinotaxis, perhaps assisted by the creation of differential flow rates in the two sides of the nose. Snakes have a very low internarial-to-skull width ratio (Fig. 1.13a), but quite widely separated nostrils. Although the snakes are larger (mean body weight 238 g) than the lizards depicted in Fig. 1.13b (mean body weight 146 g), the separation of their nostrils is very much greater. There are several possibilities for this; for example, the loose hinging of the skull which allows snakes to consume whole large prey items may necessitate that the nasal passages are forced apart to the edges of the snout, or the position of the paired Jacobson's organs may cause a displacement of the passages, but neither of these arguments *necessarily* involves the external nares. The more likely explanation is that klinotaxis adversely influences prey location and killing behaviour which, in snakes other than the heat-perceiving members of the Boidae, Crotalidae and Elapidae, depends upon vision and the use of Jacobson's organ (Noble 1936). The other anomalous group is the tube-nosed bats. Tubular nostrils have evolved twice in the bats, once in the Vespertilionidae sub-family Murininae and again in the Pteropidae sub-family Nyctimeninae. This latter group belongs to the huge assemblage of fruit-eating bats which are known to locate their food by its odour. Various suggestions have been made regarding the function of these structures, the most commonly reported one of which is that they operate as snorkel tubes allowing the bat to breath while burying its head in over-ripe fruit. However, like all mammals, bats must stop breathing when they swallow so a complex breathing device would

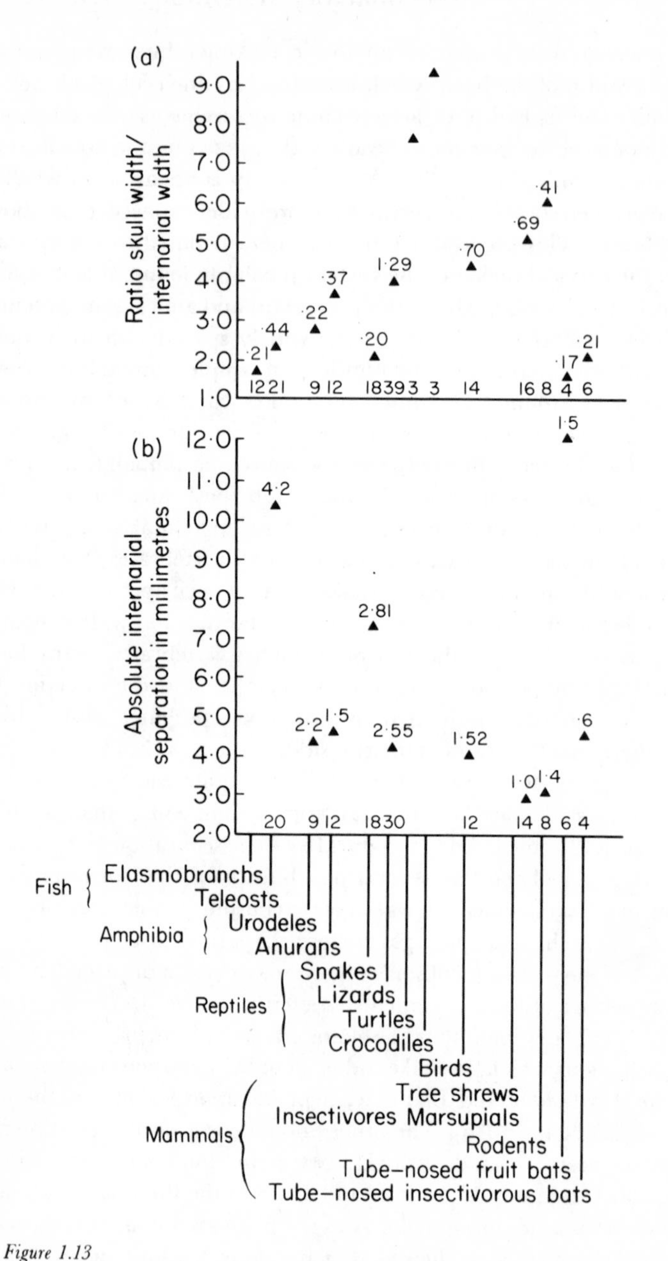

Figure 1.13
(a) The ratio of skull width to internarial width for a number of vertebrate genera.
(b) The actual internarial separation in some small genera of vertebrates
(In both (a) and (b) the small numerals above the abscissa denote the number of specimens in the sample, and those above the data point denote 2 standard errors of the mean.)

seem to be of little help. In any case, the tube is not of simple construction like that of eels, but is a delicate and complicated scroll (Fig. 1.8*o* and 1.14). Juice would certainly run into the nose through the scroll and thereby defeat the object of the snorkel. It is much more likely that the tubes allow the bat to utilize tropotactic orientation to odour while in flight, for a lateral head wagging would interfere with flight dynamics. The pronounced tubes seen in *Nyctimene* (Fig. 1.14) represent the ultimate stage in the progressive lateral development of the nares seen in the Megachiroptera. Genera such as *Myonycteris* show a distinct raising of the inner edges of the nostrils so that the angle of orientation of the nares is deflected outwards to about 45°, and are intermediate between *Nyctimene* and genera lacking tubes all together, such as the flying fox, *Pteropus*. Vespertilionids use echolocation for prey location to a greater or lesser extent and

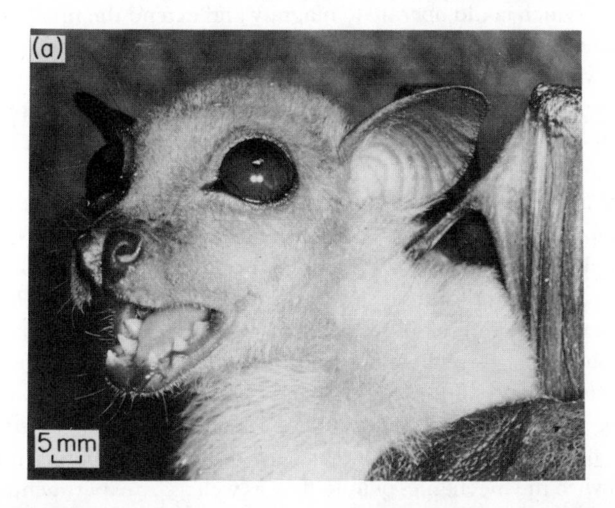

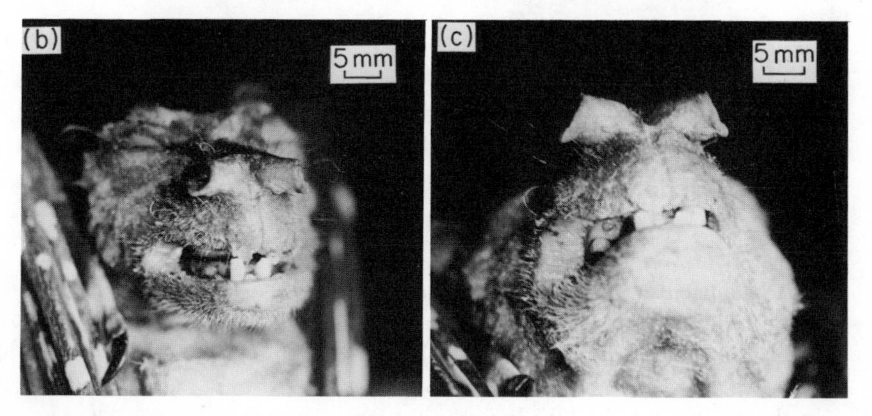

Figure 1.14 The external nares of some megachiropteran bats. (a) *Myonycteris torquata*; (b) and (c) *Nyctimene aello*. ((a) Photograph J. D. Pye.)

the same is true for the Murininae. This does not mean that olfaction plays no part in food finding; several species of insectivorous bats have been shown to discriminate between the odours of edible and distasteful species of insects, even in flight (see Chapter 2). Once again it appears likely that these small insectivorous bats benefit from tropotactic olfaction and the extension of their external nares into tubes enables them to compare incoming odours without them having to resort to klinotactic head wagging. The tubes may be used primarily or secondarily for sound beaming, but there is no evidence of this yet from these little-known bats.

In conclusion it can be said that there is no unequivocal evidence that stereolfaction is widely used by the vertebrates, although there is well-founded evidence affecting a few species. The anatomical evidence of the existence of mechanisms which could operate to magnify and extend the time-lag between a stimulus reaching first one then the other detector, or structures which effectively extend the width of the nasal part of the head, are hard to interpret in any other way. A commonly held supposition is that the *absolute* separation of the nares is the only anatomical character supporting the notion of stereolfaction, and the hammerhead shark, *Sphyrna*, is regarded as the best, and perhaps the only, example of a species equipped for this sort of olfactory perception (Fig. 1.8*l*). Such a supposition oversimplifies the complexity of an animal's sensory input and its adaptation to its ecological niche. If the point source of the odour is small, close to the animal and diffuses slowly through air or water, and the detecting animal is itself small and relatively slow-moving, a small *absolute* separation of the nares will suffice for stereolfaction. But if the odour source is large or far away from the detector and diffuses rapidly, and the detector species is both large and fast moving, a large *absolute* separation is required. This fascinating aspect of vertebrate olfactory ability is in great need of further elaboration by workers concerned with the mechanics of fluid flow as well as by experimental biologists.

1.4 Evolutionary trends

The importance of olfaction in the evolution of the vertebrates is demonstrated by the progressive development of the cerebral hemispheres of the brain. In primitive vertebrates they developed from the roof of the olfactory lobe, doubtless because olfaction was the dominant guide to movement of the entire body. As the cerebral region developed and became more complicated, the olfactory tissue spread all over the base of the brain. The name of this layer reflects its ancient origin, the palaeocortex, and it is represented in the higher vertebrates by the fornix–hippocampus complex of fibres. The olfactory bulb and its associated neurons have undergone some interesting trends as the vertebrates evolved from fish to tetrapods.

In the ancient fish, like the lamprey *Petromyzon*, undifferentiated cells carry impulses from the glomeruli in the secondary pathway. In the modern bony fishes the impulses travel along cells differentiated into two recognizable forms; mitral cells and tufted cells. Because of the manner in which these cell types interact, the transference pathways become closed or segregated. In primitive fish they remain diffuse. The second main change concerns the connection running between the bulbs in higher vertebrates. These are almost, but not completely, lacking in fish. Interbulbar connections not only link the olfactory bulbs together but they effect communication with other forebrain regions, one benefit of which might be the lowering of the threshold for transmission of incoming olfactory impulses. Other changes concern the stratification of tissue layers in the olfactory bulb. In higher vertebrates, stratification is clear and each layer serves a definite purpose concerning the correct channeling of the incoming stimulus. In fish, the lamination is poorly developed but the single bulb in the primitive hagfish, *Myxine*, is well-differentiated into clearly discernible layers (Allison, 1953). Clearly the degree of development of an organ of a species is not closely related to that species' taxonomic position.

All classes of vertebrates, with the possible exception of the birds, have a well-developed olfactory sense. In fairness it must be pointed out that a few exceptional species of birds have well-developed olfactory powers, but there are only a handful of such species. Could it be that in mastering the aerial environment and conquering a narrow ecological niche birds have lost their olfactory powers, which their Jurassic reptilian ancestors possessed and which are still seen in present-day reptiles? This is a complex question to which there is no answer. However, the similarity of structure of the avian olfactory system to that of the living reptiles and mammals makes one think they have lost the power of olfaction they once had, rather than that they never effectively gained it. The really curious thing is that the bats have partially exploited the same ecological niche as the birds but have evolved an effective and finely tuned olfactory system (Schmidt, 1975; Kämper and Schmidt, 1977). A few species of birds have converged closely with bats in their manner of food finding: south-east Asian swiftlets have evolved echolocation to assist them catch their insect prey just like the bats with whom they share cave accommodation. It is not known if there are extra ecological or behavioural reasons necessitating a well-developed sense of smell.

Wild mammals have been domesticated for man's use for countless thousands of years. Is there any evidence that removal from the natural ecological niche has brought about detectable change in the structure of the olfactory system? The answer, for the domestic pig at least, is yes. Most olfactory structures in the forebrain of domestic pigs are about 30% smaller than in wild pigs. The prepiriform complex is 35% smaller. There are other reductions however; the limbic structures of domestic pigs are up to 43% smaller than in wild pigs (Kruska and Stephan, 1973). The process of domestication involves intense selective breeding for desired characters, many of which are related to behavioural docility. It is likely, therefore, that reductions in the olfactory part of the brain are related to the obvious behavioural changes wrought by domestication.

The olfactory system of vertebrates has developed more in response to the colonization of the terrestrial rather than the aquatic environments. Airborne molecules are smaller and more volatile than those encountered in the water and, with the evolution of complex lungs, inspired air carrying these molecules has to be cleaned, warmed and moistened. So the nose and nasal apparatus of terrestrial vertebrates assumes an anatomically, as well as sensorially, enhanced importance. If it is necessary to find one's food with one's nose, one requires a good nose. The staggering development of intraspecific chemical communication among the terrestrial vertebrates, however, presents quite an interpretational problem. Most chemical signals are derived from natural cellular exudates: wastes which have been put to good use. The use of specialized chemical signals appears to have developed alongside the evolution of thermo-energetic processes and thermo-regulation (Thiessen, 1978). Increased thermo-energetic cellular activity results in a greater range of exudates being produced, by dominant individual animals, breeding animals and others experiencing a temporally induced, heightened metabolic rate. At these times, social stability is disrupted, or threatened and, only at these times, do animals communicate strongly with one another. Once equilibrium is re-established, by the intruder being expelled or by the breeding season ending, the amount of communication falls to a level at which routine activities of the population can be coordinated. Specialized chemical communication can be seen to be analogous with visual and acoustic communication in the way it operates to provide information, defuse tension and secure social harmony within populations. It is likely it has arisen in response to the active lifestyles adopted by terrestrial vertebrates, and the application of the Van 'T Hoff law governing the rate of chemical reaction in relation to temperature would release an increased number and range of metabolic wastes. It seems quite reasonable to accept that the evolutionary link between metabolic processes and specialized scent production, using cellular wastes, is strong. As far as is known, all specialized information-bearing scents are derived from wastes; in no known instance is there *de novo* production of scent.

1.5 Summary

There is evidence, taxonomic, evolutionary and structural, that the ability to perceive change in the chemical environment is as old as life itself. The chief advantage to an odour producer of the olfactory sense, compared to the visual and acoustic senses, is that a scent mark carefully deposited in the habitat can continue to broadcast its message to others long after the depositor has departed and is engaged in other activities. The sensory apparatus is held in a membrane covering the rear of the nasal passage in amphibia, reptiles, birds and mammals, and on cartilaginous leaves at the front and top of the head in fish. Only in this class is the olfactory thoroughfare not connected to the respiratory system. Jacobson's organ, the septal organ of Rudolfo-Masera, the terminal endings of the *nervus terminalis*, and the terminal endings of the *nervus trigeminus* are recognized as accessory olfactory organs, each having a discrete neural pathway to the brain.

There is little agreement about the properties of odorant molecules and various theories exist to explain odour specificity. Odorants must be reasonably volatile and at least partly soluble in both water and lipid. Odour potency increases with the increasing complexity found in homologous series.

The internarial width of mammals is, relative to the maximum width of the skull, narrower than that for sharks and bony fish. The external nares of a few species are held out from the body on a flap of tissue (hammerhead shark) or on the end of short tubes (tube-nosed bats, eels). This could be an adaptation to stereolfaction reception. Stiff-necked species, and those flexible-necked species requiring to keep their heads steady while hunting (snakes, bats), require to have their external nares as widely spaced as possible if they are to orientate by lateral odour sampling.

The use of specialized secretions for intraspecific communication appears to have evolved alongside the evolution of the high metabolic rate seen in the homoiothermic vertebrates. *De novo* production of communicatory odour has not been recorded; odorous materials are refined cellular waste products.

2

Sources and chemistry of vertebrate scent

The odorous environment of vertebrate animals comprises elements of diverse origins, from those associated with food to those associated with natal nest sites or spawning streams and which act as homing beacons for returning migrants. Each in its own way plays a fundamental and characteristic part in the behavioural ecology of the species, and such odours are just as important as those produced by the individual animal itself for purposes of bringing about and maintaining the social organization of the species. Self-produced odours are aimed almost exclusively at conspecifics (the exceptions to this are those odours produced for defensive purposes) and their production is carefully matched to their dissemination. Vertebrates show a wide range of scent deposition behaviours, each one designed to place the odorous secretion in that part of the environment where it will be most noticed by conspecifics. Most specialized behaviours are derived from behaviours associated with elimination, i.e., urination and defaecation, though the application of scent secretions from some specialized sebaceous glands requires the acquisition of new behaviour patterns. There are many reviews of scent-marking behaviour available, and the reader should consult one of these for details (Ralls, 1971; Johnson, 1973; Stoddart, 1976a). In general, it appears that active scent setting is a mammalian phenomenon; in snakes and lizards scent which is produced by the dorsal and cloacal glands disperses passively as the animals travel through the vegetation. By way of a brief summary of scent-marking behaviour, Table 2.1 indicates the main scent-marking behaviour and passive marking pathways in mammals.

This chapter investigates the sites of production of vertebrate odours and their chemical compositions; it also considers the threshold concentrations necessary to invoke neural and behavioural responses.

Table 2.1 Active and passive marking behaviour and pathways in mammals. Passive mark pathways exclude the influence of rectal glands on faeces, the odour of urine, and the odour of saliva left on partially eaten foods.

Order	Active marking behaviour patterns	Passive mark pathway
Monotremata	Anal drag	
Marsupialia	Cloacal drag; sternal gland, frontal gland and chest rub; salivation	
Insectivora	Salivation	Lateral gland rub
Chiroptera	Largely unknown; submandibular scent glands; sac-like gland in wing membrane of *Sacopteryx*	
Primates	Urination on hands and feet; punctuated micturition; dribble urination; sternal gland, gular and brachial gland rub on conspecifics, branches and on tail which is waved before opponent; chinning in tree shrews; anal drag in marmosets	
Edentata	Unknown	
Pholidota	Anal drag	
Lagomorpha	Chinning	
Rodentia	Anal drag; urination on ground; salivation; scratching of lateral glands	Lateral gland rub; dorsal gland rub; palmar gland press
Cetacea	Presumed absent	
Carnivora	Urination, leg cocking by males occasionally by females; ano-genital drag; ventral drag; handstand anal drag.	
Tubulidentata	Unknown	
Proboscidea	Temporal gland rub, or transference via trunk	
Hyracoidea	Dorsal gland rub	
Sirenia	Presumed absent	
Peissodactyla	Ritualized defaecation and urination (*Diceros*)	
Artiodactyla	Ritualized defaecation and urination (*Hippopotamus*); urine rub; pre-orbital facial gland rub; abdominal tail gland rub; body rub; tarsal gland rub	Interdigital gland press

2.1 The sites of odour production

Vertebrates utilize a great many anatomical sites and physiological processes in the production of odours. All exudative processes have been implicated in odour production, and urine and faeces, laced with preputial and anal gland secretions, may be the most important odour vehicles for many species. A characteristic of the mammals is their soft yet waterproof skin, well-endowed with glands secreting substances whose primary function is to maintain skin and hair condition (Nicolaides, 1974). In very many, if not all, types of mammal, certain areas of skin have taken over a secondary role of producing odorous substances, and the normal sebaceous and sudoriferous glands show enormous hypertrophy correspondingly. Products of the salivary glands are of great importance in some species, notably in pigs, *Sus scrofa*, in which saliva odorants rigorously control courtship and mating behaviour (Patterson, 1968). Autosalivation occurs in several species of mammal, including the hedgehog, *Erinaceus europaeus* (Brockie, 1976). Self-anointment with saliva occurs only during the breeding season and is frequently elicited by the presence of hedgehogs of the opposite sex. Although self-anointment has only been observed to occur in the wild during the breeding season, in the laboratory it can be induced by novel situations in any month of the year. As it is not invariably associated with mating in the wild, it may be nothing more than a piece of redirected behaviour borne of the conflict of behaviours seen in courtship. Autosalivation occurs in dogs and kangaroos for thermoregulatory purposes, but this would not seem to be its purpose in hedgehogs.

In several species of primate the presence of carboxylic acids in the vaginal discharge acts strongly to attract the attention of males. These attractants, which have been named 'copulins' (Michael and Bonsall, 1977), are the products of bacterial decay of glandular secretions. The effect of washing the urino-genital area, and the removal not only of the copulins but also the seminal fluid remaining from a previous mating, is to reduce the attractiveness of the female. In sheep and goats, and probably a great many other species as well, the bond necessary for the expression of full maternal behaviour is forged by the dam licking and sniffing some of the birth fluids surrounding the youngster (Baldwin and Shillito, 1974).

Although rather little is known about the active ingredients in these secretions outlined above, there is little evidence to suggest that the behavioural interest has arisen in respect to anything other than substances primarily produced for their apparent purpose. Any olfactory stimulation is seen to be of secondary value, though of undoubted importance. The products of discrete scent glands, however, have no function other than as sources of odorants. This is their primary purpose and behaviours relating to individuals donating scent or perceiving scent are primary olfactory behaviours. Table 2.2 summarizes such glandular development and indicates the reported olfactory role of each. It must be stressed that the list of behaviour roles includes only those exemplified by distinctive, overt behaviour and ignores any priming effects of odour, i.e., when a particular odour influences behaviour via an indirect action on the endocrine and other metabolic systems and the central nervous system (Wilson and Bossert, 1963). It is thus a great underestimation. Although the range of anatomical sites used for odour production is high, it can be stated that those associated with the genitalia relate to reproductive behaviour, while those sited on the ventrum, dorsum, head and flank region relate to territorial and other behaviours associated with social status and property tenure.

Table 2.2 Summary of the type and position of specialized scent glands found in the vertebrates.

Class	Sub-class taxon	Genus or species or family	Type and position of odorous glands	Reported olfactory role	Reference
Pisces		Ostariophysi and Gonorhynchi	Dermal club cells	alarm	von Frisch 1938
Amphibia	Caudata	*Plethodon* and other salamanders	Mucus cells, chin	courtship	Madison, 1977
			Cloacal, pelvic and abdominal dermal	courtship	
	Anura	*Rana*, *Bufo*, *Gastrotheca*, *Phyllomedusa*	Granular and mucus dermal	alarm; defence	Noble, 1931 Baylock, Ruibal and Platt-Aloia, 1976
Reptilia	Lacertilia		Pre-anal, epidermal follicular structures	communication	Maderson, 1970
		Crotaphytus collaris	Femoral sebaceous cells	courtship?	Cole, 1966
		Scleropus graciosus	Cloacal subcutaneous apocrine	territorial?	Burkholder, and Tanner, 1974
	Ophidia		Cloacal holocrine	alarm; species, individual and sex recognition. courtship	Noble and Clausen, 1936
		Natrix Macropisthodon	Nucho-dorsal holocrine	alarm; courtship?	Smith, 1938
	Chelonia	Testudinidae	Chin holocrine	courtship; combat	Winokur and Legler 1975 Rose, 1970
		turtles	Inguinal and axillary holocrine	courtship; defence	Ehrenfeld and Ehrenfeld, 1973
	Crocodilia		Mandibular; cloacal holocrine and apocrine	courtship; combat	Guggisberg, 1972
Aves	all orders		Uropygidial holocrine	courtship?	Paris, 1914

Table 2.2 Continued

Class	Sub-class taxon	Genus or species or family	Type and position of odorous glands	Reported olfactory role	Reference
Mammalia	Marsupialia		Sternal, ventral, gular, frontal holocrine	territorial? group and individual recognition	Ewer, 1968 Schultze-Westrum 1965
	Chiroptera		Sternal, scapular, submandibular	individual recognition, territorial sex attraction	Nelson, 1965 Allen, 1941
	Primates	Cebidae, Hapalidae, Indridae, Lemuridae, Lorisidae, Tupaiidae, Callithricidae	Sternal, gular, brachial sebaceous apocrine	territorial, object marking, establishment of social status	Thiessen and Rice, 1976
	Edentata	*Bradypus*	Mid-dorsal epidermal gland	object marking	Tembrock, 1968
	Pholidota	*Manis*	Ano-genital sebaceous	territorial reassurance, advertisement of sexual condition	Pagès, 1968
	Lagomorpha	*Oryctolagus cuniculus*	Chin, anal, inguinal sebaceous-apocrine	territorial reassurance, group and individual recognition	Mykytowycz, 1968
	Rodentia	Cricetidae, Heteromyidae, Muridae, Sciuroidea	Ventral flank, hip, supra-caudal sub-caudal, anal, chin, Meibomian sebaceous	object marking; sex, individual and group recognition	Thiessen and Rice, 1976
	Carnivora	Canidae, Viverridae, Mustelidae	Supra-caudal, ventral, facial, anal	object marking; territorial	Thiessen and Rice, 1976 Ewer, 1968
	Proboscidea	Elephantidae	Temporal apocrine	courtship?	Kühme, 1961 Eisenberg *et al.*, 1971
	Hyracoidea		Dorsal holocrine	alarm?	Schaffer, 1940

Artiodactyla	Cervidae	Pre-orbital, tarsal, metatarsal, interdigital	object marking, individual and sex recognition;	Hatlapa, 1977 Müller-Schwarze, 1971
	Bovidae	Pre-orbital, occipital, supra-caudal, inter-mandibular apocrine	object marking; expression of social status; alarm	Hediger, 1949 Cohen and Gerneke, 1976 Millais, 1895
	Antilocapridae	Sub-auricular, rump	object marking; alarm	Müller-Schwarze *et al.*, 1974 Seton, 1927
	Tylopodidae	Temporal holocrine and apocrine	courtship	Pilters, 1956
		Dorsal	courtship, territorial; object marking	Sowls, 1974 Frädich, 1967

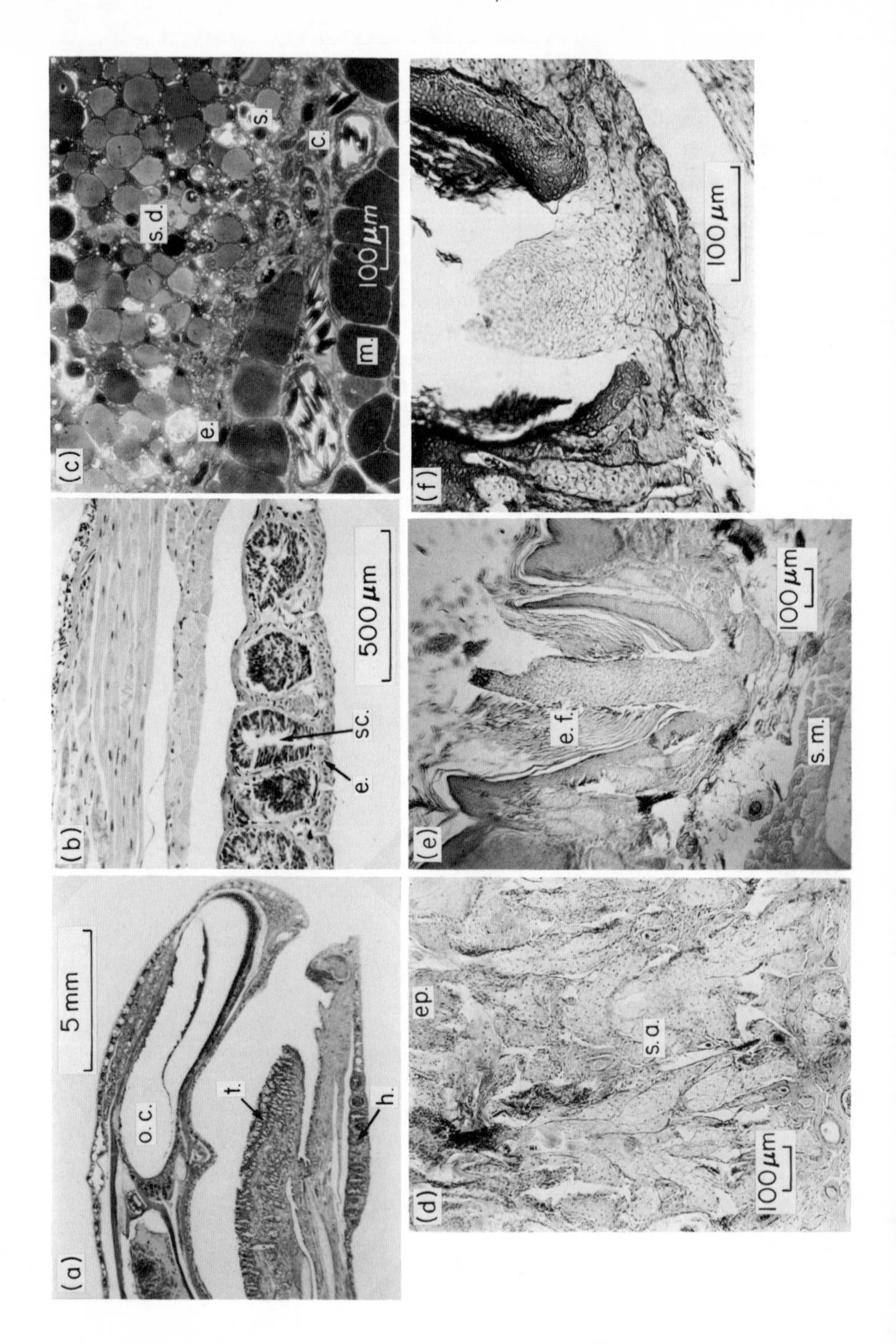
(a)
5 mm
o.c.
t.
h.
(b)
sc.
e.
500 μm
(c)
s.d.
s.
c.
e.
m.
100 μm
(d)
ep.
s.a.
100 μm
(e)
e.f.
s.m.
100 μm
(f)
100 μm

Figure 2.1

There is surprisingly little variation in the structure of scent-producing glands found in reptiles and mammals. Mostly such glands consist of hypertrophied normal skin elements and usually are associated with a thickened epidermis. This sloughs off in small fragments and mixes with the secretion, and possibly acts as a granular 'filler' having a physical, rather than a chemical function. The structure of typical amphibian, reptile and mammalian odorous glands are shown in Fig. 2.1. Broadly speaking, the sebaceous acini in these glands are greatly hypertrophied and, in some mammals, their associated hair follicles have disappeared. The peripheral region of the acinus is characterized by much active mitosis in holocrine glands with a constant migration of new cells to replace those expressed as secretion. Cutaneous glands are heavily influenced by the sex hormones, testosterone and oestrogen. Testosterone and other androgens increase both cell size and cell proliferation and generally stimulate glandular activity. Oestrogens exert a broadly inhibitory effect by decreasing the rate of passage of cells through the gland. In this way the size of the gland is reduced, but the main inhibitor to sebum production is the quelling effect of oestrogens on cell proliferation (Ebling, 1963, Strauss and Ebling, 1970).

Seasonally breeding species which bear specialized sebaceous gland complexes sometimes show massive glandular recrudescence at the start of breeding. Fig. 2.2 shows the pattern of seasonal waxing and waning of the flank organ of the rodent *Arvicola terrestris* (Stoddart, unpublished). Histological sections of the organ show that during the early breeding season the surface rises evenly in response to the pressure of the enlarging sebaceous acini from beneath (Fig. 2.3a). At the height of the breeding season the surface starts to become slightly irregular as the secretory acini start to degenerate (Fig. 2.3b), and at the end of the breeding season the degeneration is complete with the epidermis collapsed such that the surface of the gland becomes concave (Fig. 2.3c). In Britain, where

---

Figure 2.1 Histological structure of the scent-producing organs of an amphibian, a reptile and a mammal.

(a) Longitudinal section through head of the newt, *Plethodon jordani*. h. hedonic scent gland; o.c. olfactory cavity; t. tongue. (Photograph: D. M. Madison)

(b) High-power view of secretory cells in the hedonic gland of *P. jordani*. s.c. secretory cell; e. epidermis. (Photograph: D. M. Madison)

(c) Section through Rathke's gland of the turtle, *Sternotherus odoratus*. s. secretory cell; e. epithelial cell; s.d. secretion droplet; m. striated muscle; c. connective tissue. (Photograph: J. G. Ehrenfeld)

(d) Transverse section through flank organ of adult male rodent, *Arvicola terrestris*, showing the massive hypertrophy of the sebaceous acini. s.a. sebaceous acini; ep. epidermis.

(e) Pit formed in the surface of flank organ of *A. terrestris*, showing an expressed pillar of secretion. Note the sloughed fragments of keratinized epidermis. e.f. epidermal fragments; s.m. subcutaneous muscle.

(f) High-power view of an epidermal pit in the flank organ of male *A. terrestris*, to illustrate the holocrine nature of the secretion.

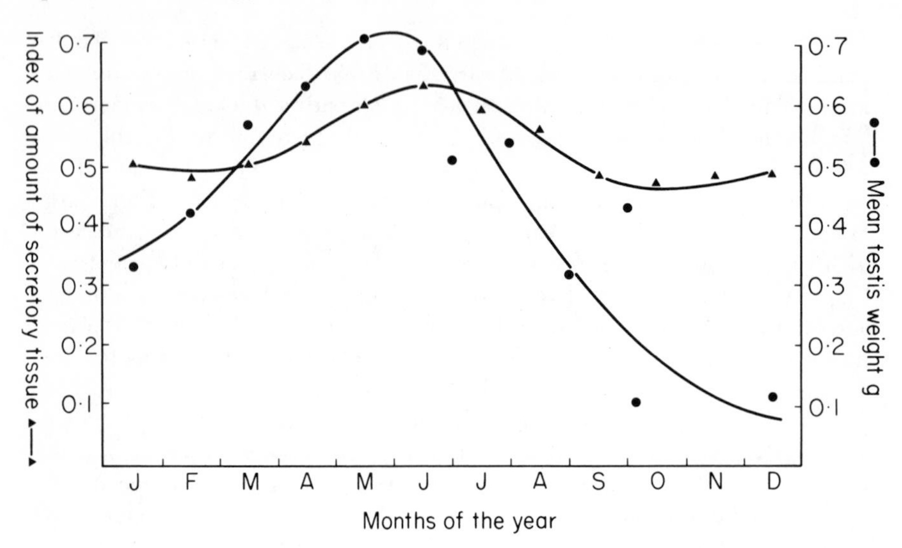

Figure 2.2 Seasonal cycle of secretion production by flank organ of adult male *Arvicola terrestris*. (From author's own unpublished data.)

the winter climate is mild, the process of redevelopment of the organ starts during December. By contrast, the dorsal sebaceous gland of the North American kangaroo mice, *Dipodomys*, shows a marked cycle of activity in some species but not in others, yet all breed on a seasonal basis (Quay, 1953). Observations such as these indicate that there may be more than a single function for the secretions of those glands which respond to the changing blood levels of sex hormones.

Detailed studies on the chemical constitution of the flank organ secretions of *Arvicola terrestris* indicate that while the amount of secretion is considerably reduced during the period of reproductive quiescence, the composition of that which is produced is quantitatively different from that produced during the breeding season (Stoddart, Aplin and Wood, 1975). It appears that the same substances are present throughout the year, but the proportional amount of each varies according to the time of year. This is shown in Fig. 2.4 where the relative amount of just one group of compounds in the flank organ secretion of one individual male *A. terrestris* is expressed as a measure of how much it varies from its level during the height of the breeding season. The analysis was performed on a frequently retrapped member of a wild population and the figure shows that the maximum amount of change occurs around the time of onset and cessation of breeding. Only one group of compounds is shown, but analysis of the pattern of occurrence of many groups reveals that all undergo a substantial quantitative change in their relative amounts at these two times of the year (Tomkins, unpublished data).

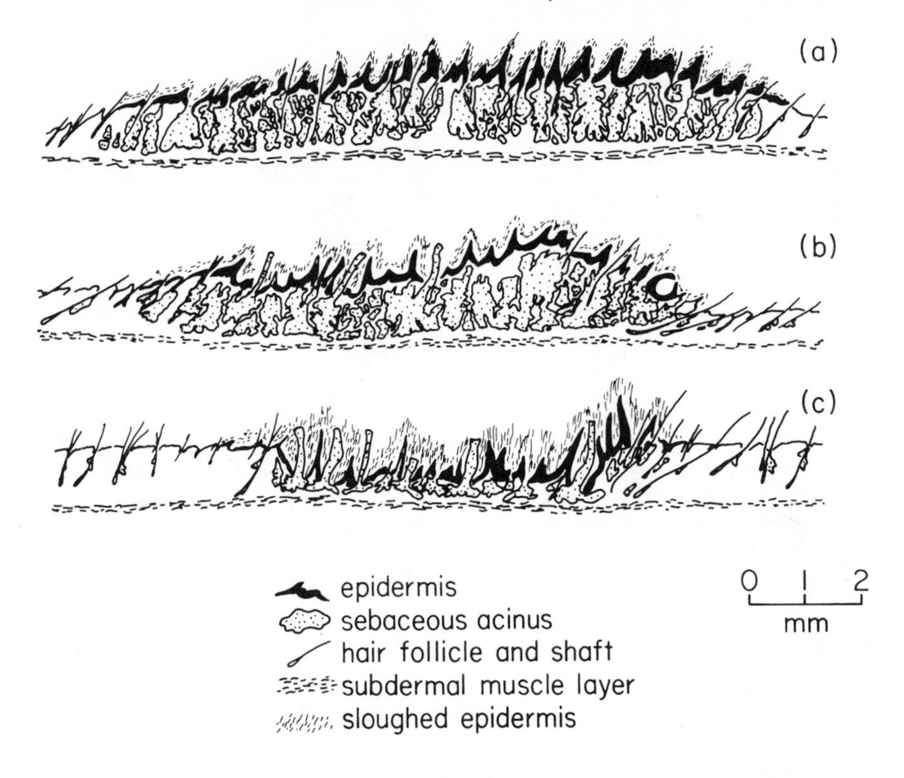

Figure 2.3 Stylized transverse sections through the flank organ of adult male *Arvicola terrestris*.
(a) early breeding season (March),
(b) middle of breeding season (July),
(c) late breeding season (October).
Note the collapse of the surface of the organ and the associated production of thick pillars of secretion. (Redrawn from Vrtiš, 1930, with considerable modification from the author's own unpublished observations.)

Many glandular structures of mammals are hairless or almost so (see Fig. 2.3), but by no means all. Osmetrichia, or specialized 'odour hairs' have been described for a few species of mammals, but they may be very common. Hairs protruding from the ventral gland of the Mongolian gerbil (*Meriones unguiculatus*) are grooved and point towards the rear (Fig. 2.5). This arrangement allows the best possible deposition of sebum as the animal moves forward over an object (Thiessen and Rice, 1976). The cuticular scales on the hairs of the tarsal organ of black-tailed deer, *Odocoileus hemionus*, are specially designed to enable them to carry glandular and urinary compounds for longer than would otherwise be the case (Fig. 2.5) (Müller-Schwarze, Volkman and Zemanek, 1977). A series of comb-like ridges pointing upwards and backwards from the edge of the scale

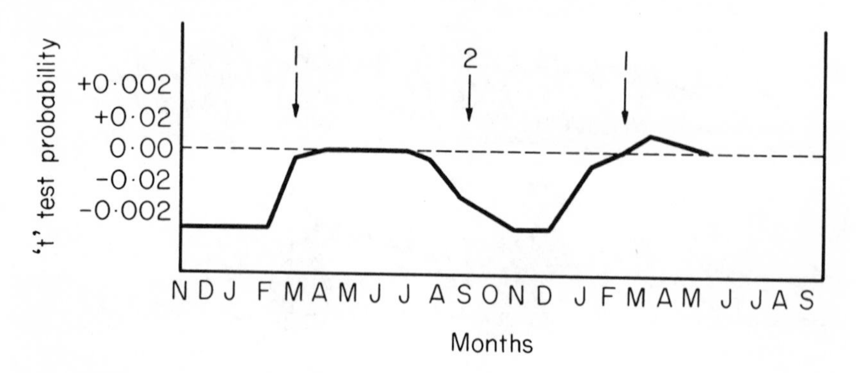

Figure 2.4 The amount of change, measured by a 't' test probability, in the relative amount of one of the groups of compounds present in the whole flank organ secretion of *Arvicola terrestris* and observed over a period of 18 months. The 0.00 line is the mean relative value for this group of compounds occurring in the middle of the breeding season (June). The numbered arrows denote the start of breeding (1) and the finish of breeding (2). The data refer to one individual male living in a wild population. (Unpublished data by courtesy of G. W. O. Tomkins.)

combines with large chambers beneath the scales to give the hair the appearance of a long, ripe pine cone. The short white hairs lining the lateral gland of the crested rat, *Lophiomys imhausi*, have an even more curious structure (Figs. 7.3 and 7.4). Each hair, which measures up to 2.5 cm in length, has a vacuolated interior resembling that of a loofah body brush, which is capable of holding a volume of secretion about four times the volume of the hair. Further consideration of the olfactory biology of *Lophiomys* is delayed until Chapter 7.

2.2 Chemical composition of odorants

The range of chemical substances manufactured by scent organs of vertebrates appears to be almost limitless. Table 2.3 displays a summary of what is known about the chemistry of the secretions of specialized superficial scent glands.

Figure 2.5
(a) and (b) Specialized hair from the centre of the tarsal tuft of a yearling female black-tailed deer, *Odocoileus hemionus columbianus*. Note the ridges on the edge of the cuticular scale and spaces between the scales. (a × 350; b × 975)
(c) Unspecialized hair from the back of the same yearling female (× 230).
(d) and (e) Hairs from the ventral gland of an adult male Mongolian gerbil, *Meriones unguiculatus*. Note the deep channel for the transmission of the secretion. (d × 150; e × 455).
(Reproduced from Müller-Schwarze *et al.*, 1977, by kind permission of Academic Press.)

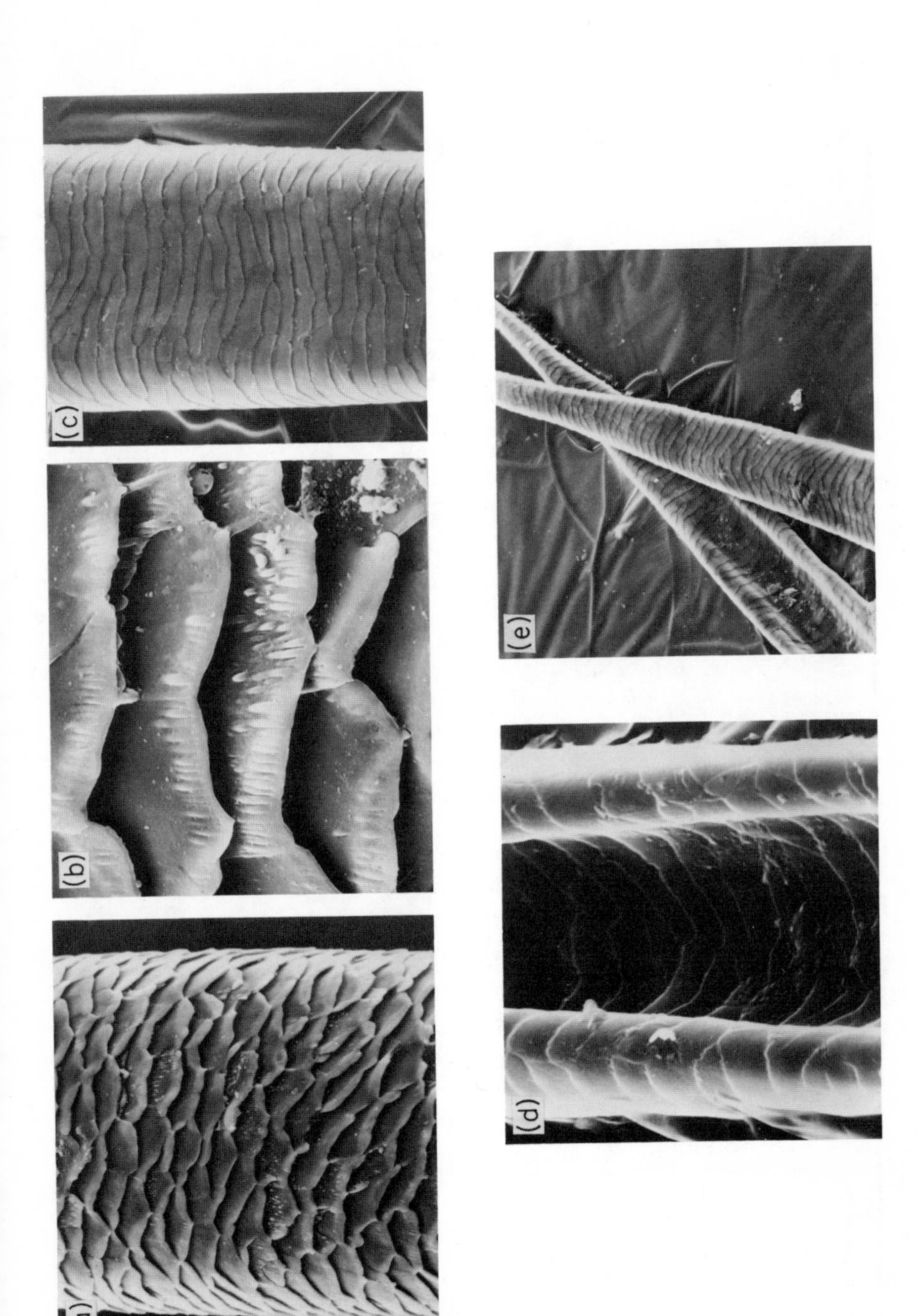

Figure 2.5

Table 2.3 The chemistry and behavioural significance of vertebrate odours.

Genus species	Type of gland	Seasonal cycle of activity	Presence in sexes ♂	♀	Chemical composition of secretion	Determination technique	Behavioural significance	Reference
1	2	3	4		5	6	7	8
PISCES								
Bathygobius soporator (blind goby)	Ovary	√		√			Courtship*	Tavolga, 1956
Blennius pavo (blenny)	Cutaneous anal		√				Attraction of female*	Eggert, 1931
Ictalurus catus (white catfish)	Cutaneous club cells	×	√	√	Amino acids or similar compounds		Alarm reaction*	Tucker and Suzuki, 1972
Clinostomus funduloides (rosyside dace)	Cutaneous club cells				Histamine, small ringed or double-ringed compounds		Alarm reaction*	Reed *et al.*, 1972
Notropis cornutus (common shiner)	Cutaneous club cells				Histamine, small ringed or double-ringed compounds		Alarm reaction*	Reed *et al.*, 1972
Phoxinus phoxinus (European minnow)	Cutaneous club cells	×	√	√	A pterin, conjugated with protein		Alarm reaction*	Pfeiffer and Lemke, 1973

AMPHIBIA								
Bufo bufo (common toad)	Cutaneous	×	√	√	Bufotoxin; γ-bufotoxin		Alarm reaction*	Pfeiffer, 1963
REPTILIA								
Gopherus berlanderi (tortoise)	Holocrine submandibular	√	√	√	Caprylic, capric, lauric, myristic, palmitic, palmitoleic stearic, oleic, linoleic acids		Courtship, combat	Rose, 1970
Sternotherus odoratus (stinkpot turtle)	Cutaneous	×	√	√	Phenylacetic, 3-phenylpropionic 5-phenylpentanoic, 7-phenylheptanoic acid		Alarm reaction*	Eisner *et al.*, 1977
Alligator sclerops (alligator)	Suboptical and throat glands				*Iso*-valeric, myristic and palmitic acids, acetyl alcohol. Also 'yacarol' – mainly *d*-citronellol	1	Courtship	Lederer, 1950
Colubrid, boid and viperid snakes	Holocrine cloacal	×	√	√	Complex series of lipids	2	Inter- and intraspecific recognition	Oldak, 1976
AVES								
All species	Holocrine uropygidial	√	√	√	2, 3-diols (2, 4, 6, 8-methyl decanoate and 2, 4, 6, 8-methyl undecanoate in the goose)			Nicolaides, 1974; Paris, 1914
MAMMALIA								
Viverra civetta (African civet)	Anal sac		√		Cyclohexadecanone, cycloheptadecanone, 6 *cis*-cycloheptadecenone, 9 *cis*-cycloheptadecenone, 9 *cis*-cyclononadecenone			van Dorp *et al.*, 1973

Table 2.3 Continued

Genus species	Type of gland	Seasonal cycle of activity	Presence in sexes ♂	♀	Chemical composition of secretion	Determination technique	Behavioural significance	Reference
1	2	3	4		5	6	7	8
MAMMALIA								
Hyaena hyaena (hyaena)	Anal sac		√	√	5-thiomethylpentane-2, 3-dione		Individual recognition	Wheeler *et al.*, 1975
Canis familiaris (domestic dog) and *C. latrans* (coyote)	Anal sac	×	√	√	Carboxylic acids C_2-C_6, 2-piperidone, aldehydes, trimethylamine	3, 4	Range marking*	Preti *et al.*, 1976
Vulpes vulpes (fox)	Supracaudal Anal sac		√		Dihydroactinidiolide Saturated carboxylic acids C_2-C_6, putrescine (1, 4, diaminobutane), trimethylamine, cadaverine (1, 5, diaminopentane), 5-aminovaleric acid, indole, occasionally stanol (5β-cholestan 3β-ol)	2, 3, 4	Territory marking; alarm signal	Albone, 1975 Albone and Perry, 1976
Panthera leo (lion)	Anal sac		√		Saturated carboxylic acids C_2-C_6, phenylacetic, 3-phenylpropionic, *p*-hydroxyphenylacetic and (*p*-hydroxyphenyl) propionic acid.	2, 3, 4		Albone and Eglinton, 1974

					Homologous series of 1-alkylglycerols and 2-hydroxy fatty acids and 3-hydroxy fatty acids			
Panthera tigris (tiger)	Ejected with urine			√	Primary amines	1	Expression of sexual state	Brahmachary *pers. comm.*
Mustela putorius (polecat)	Anal gland		√	√	Mustelan ($C_5H_{10}S$), thietane, 3, 3-dimethyl-1, 2-dithiolane, di-isopentyl disulphide	1, 3, 4		Schildknecht *et al.*, 1976
Mustela vison (Canadian mink)	Anal gland secretion		√	√	Acetic, propionic, *n*- and *iso*-butyric, *iso*-valeric, α-methylbutyric, *n*-valeric, *iso*-caprionic acids, proteins, carbohydrates, saturated and unsaturated lipid compounds, hydrocarbon, sulphur compounds, monoesters, carbonyl compounds, alcohols, nitrogen compounds.	1, 3, 4, 5, 6		Schildknecht *et al.*, 1976; Zinkevitch, *pers. comm.*
	Anal gland headspace analysis				3, 3-dimethyl-1, 2-dithiolane, 2, 2-dimethylthietane	3, 4, 5, 6		Zinkevitch, *pers.comm.*
Mustela vison (Canadian mink)	Vaginal secretion	√		√	Methylamine, di- and trimethylamine, ethylamine, di- and triethylamine, propylamine, dipropylamine, butylamine, pyridine, α-picoline, piperidine, n-methylpyrrole		Recognition of sexual status	Sokolov and Khorlina, 1976 Sokolov *et al.*, 1974
Mustela nivalis (weasel)	Anal gland		√	√	Mustelan, indole, $C_{10}H_{20}$ carbohydrates	1, 3, 4		Schildknecht *et al.*, 1976

Table 2.3 Continued

Genus species	Type of gland	Seasonal cycle of activity	Presence in sexes ♂	Presence in sexes ♀	Chemical composition of secretion	Determination technique	Behavioural significance	Reference
1	2	3	4		5	6	7	8
MAMMALIA								
Mephitis mephitis (skunk)	Anal sac		√	√	*trans*-2-butene-1-thiol, 3-methyl-1-butanethiol, *trans*-2-butenyl methyl disulphide, di-butyl mercaptan	3, 5, 6	Defence*	Andersen and Bernstein, 1975
Herpestes auropunctatus (Indian mongoose)	Anal sac		√	√	Saturated carboxylic acids C_2-C_6	3	Object marking,* individual recognition*	Gorman *et al.*, 1974
Procavia capensis (hyrax)	Cutaneous dorsal		√	√	A paraffin, a saturated alcohol C_{20}, an unsaturated alcohol $C_{26}H_{54}O$, palmitic, stearic, oleinolic and ursolic acid, a mixture of sterols and three triterpenes	1	Alarm; warning?	Lederer, 1950
Odocoileus dichotomus (marsh deer)	Nasal	×	√	√	Cholesterol esters, mono- and di-ester waxes (containing 2-hydroxy fatty acids). Alcohols of the di-ester waxes contain mono-unsaturated homologues with double bonds in (ω-9) position.	3, 4		Jacob and v. Lehmann, 1976

Odocoileus hemionus (mule deer)	Tarsal	(✓)	✓	*cis*-4-hydroxydodec-6-enoic acid lactone		Individual recognition*	Brownlee *et al.*, 1969
Antilocapra americana (pronghorn antelope)	Subauricular	(✓)		*Iso*-valeric, 2-methylbutyric, 13-methyl-1-tetradecanol, 12-methyl-1-tetradecanol, 13-methyl-tetradecyl 3-methyl-butyrate, 12-methyltetradecyl 3-methyl-butyrate, 13-methyltetradecyl 2-methyl-butyrate, 12-methyltetradecyl 2-methylbutyrate	3, 4, 5, 6	Object marking*	Müller-Schwarze *et al.*, 1974
Rangifer tarandus (reindeer)	Tarsal	✓	(✓)	*n*-heptanol, *n*-octonol, *n*-nonanol, *n*-decanol, *n*-dodecanol, *n*-tetradecanol, *n*-hexadecanol, *n*-heptadecane. (Valeric and *iso*-valeric acids additionally present in spring) *n*-octadecanol-1, *n*-hexanol-1. Lanestrol, cholesterol esters, fatty acids, trigycerides and higher fatty acids	3, 4	Individual and sex recognition	Andersson *et al.*, 1975; Müller-Schwarze *et al.*, 1977; Sokolov *et al.*, 1977
Rangifer tarandus (reindeer)	Interdigital	✓	✓	carboxylic acids C_2-C_6, lanesterol, cholesterol, triglycerides, (C_{12}-C_{20} sterol esters as unhydrolysed precursors)	2, 3, 4, 5		Brundin *et al.*, 1978
	Caudal	✓	✓	*n*-heptanal, *n*-octanal, acetic acid, *n*-nonanal, propionic acid, *iso*-butyric acid, *n*-decanal, *n*-butyric acid, *iso*-valeric acid, methyl butyl acid	3, 4		

Table 2.3 Continued

Genus species	Type of gland	Seasonal cycle of activity	Presence in sexes ♂	♀	Chemical composition of secretion	Determination technique	Behavioural significance	Reference
1	2	3	4		5	6	7	8
MAMMALIA								
Damaliscus dorcas (blesbok)	Pedal				5-undecen-2-one, 2-heptanone, 2-nonanone, 2-undecanone, 2, 5-undecanedione, α-terpineol, 2-n-heptylpyridine, *m*-cresol, (2)-6-dedecen-4-olide	3, 4, 5, 6	Range marking*	Burger *et al.*, 1976; Burger *et al.*, 1977
Moschus moschiferus (musk deer)	Ventral pouch		√	√	9 *cis*-cycloheptadecenone			Lederer, 1950
Loxodonta africana (African elephant)	Temporal gland		√	√	*m*- and *p*-cresol, phenol, indole	3, 4	Expression of stress; warning	Adams *et al.*, 1978
Sus scrofa (pig)	Submaxillary		(√)		5α-androst-16-en-3-one, 3α-hydroxy-5α-androst-16-ene	3, 4	Induces lordosis in sows*	Perry *et al.*, 1973
Sanguinus fuscicollis (brown-headed tamarin)	Circum genital	?	√	√	Series of saturated butyrates	3, 4, 5	Individual recognition	Yarger *et al.*, 1977
	Sternal				Mono-unsaturated butyrates of two structural types		Sex recognition	

				Di-unsaturated butyrates			
	Urine, urine plus vaginal secretions						
Macaca mulatta (macaque)	Vaginal secretion	✓	(✓)	Acetic, propionic, methyl-propionic, butanoic, 3, me-thylbutanoic, 4, methylpen-tanoic acids	3, 4	Expression of sexual status*	Michael and Bonsall, 1977
Homo sapiens (man)	Vaginal secretion	✓	(✓)	C_3-C_6 aliphatic acids, acetal-dehyde, ethylformate, ac-etone, ethylacetate, ethanol, 1, 3-dioxolane, 3-methyl-1-butanol, 2-5-dihydrofuran, benzamide or benzonitrile, furfuryl alcohol	3, 4		Michael *et al.*, 1974; Keith *et al.*, 1975; Huggins and Preti, 1976
	Axillary	✓	✓	5α-androst-16-en-3α-ol			Brooksbank *et al.* 1974
	Generalized cutaneous	✓	✓	Squalene, wax esters, triacyl glycerols, di- and monoacyl glycerols, unesterified fatty acids, sterol esters			Nicolaides, 1974
Papio anubis (sacred baboon)	Vaginal secretion	✓	(✓)	C_3-C_6 aliphatic acids		Expression of sexual status	Michael *et al.*, 1972
Galago sp. (bush baby)	Ventral		(✓)	2-(*p*-hydroxyphenyl) ethanol			Wheeler *et al.*, 1975
Oryctolagus cuniculus (rabbit)	Anal		(✓) ✓	6-methylhept-5-enal, 6-methyloctanal, 7-methyl-nonanal, 8-methylnonanal, *n*-decanal, non-2-enal, dec-2-enal, *cis*-undec-4-enal, *cis*-dodec-5-enal, *trans*-dodec-5-enal, *n*-dodecanal, *n*-nonanal, *n*-undecanal	3, 4, 5	Decelerates heart rate in males*	Goodrich *et al.*, 1978

Table 2.3 Continued

Genus species	Type of gland	Seasonal cycle of activity	Presence in sexes ♂	Presence in sexes ♀	Chemical composition of secretion	Determination technique	Behavioural significance	Reference
1	2	3	4		5	6	7	8
MAMMALIA								
	Chin		√	√	Tri-, di-, and monoglycerides		Territorial marking* and marking of young.*	Goodrich and Mykytowycz, 1972
	Inguinal		√	√	Tri-, di-, and monoglycerides			
Castor fiber (beaver)	Anal	√	√	√	Benzyl alcohol (free and esterified), β-cholestanol, mannitol, *cis*-5-hydroxytetrahydroquinol, *p*-ethylphenol, *p*-propylphenol, pyrocatechol, quinol, quinolmonomethyl ether, chavicol, ethyl anisol, 4 methyl and 4 ethyl pyrocatechol, betuliginal, 2, 4-dihydroxydiphenylmethane, 2, 3-dihydroxydibenz-2-pyrone, 4, 4-dihydroxydiphenic acid dilactone, salicaldehyde, castoramine, acetophenone, *p*-hydroxyacetophenone, *p*-methoxyacetophenone, benzoic acid, β-phenyl-		Territorial marking*	Lederer, 1950

					propionacetic acid, cinnamic acid, salicylic acid, *m*-hydroxybenzoic acid, *p*-hydroxybenzoic acid, anisic acid, genistic acid, 5-methoxysalicylic acid, a phenolic ester, esters of ceryl and benzyl alcohol, esters of phenols and genistic acid.			
Mesocricetus auratus (golden hamster)	Vaginal secretion	√	(√)		Dimethyl disulphide	3, 4	Sex attractant*	Singer *et al.*, 1976
Meriones unguiculatus (Mongolian gerbil)	Sternal	×	(√)	√	Phenylacetic acid. Cholesta-3, 5-diene in males only	3, 4	Territorial marking*	Thiessen *et al.*, 1974; Jacob, 1977
Galea musteloides (cuis)	Chin		(√)		Various steroid metabolites	1	Courtship, sexual display	Holt and Tam, 1973
Arvicola terrestris (water vole)	Flank	√	(√)	√	Long chain esters of long chain alcohols (acids C -C; alcohols C -C), triglycerides	2, 3, 4	Range marking*	Stoddart *et al.*, 1975
Mus musculus (mouse)	Preputial		(√)		Wax ester, neutral plasmalogen, glyceryl ether diester and triglyceride.	2, 3, 5		Sansone and Hamilton, 1969
Ondatra zibethicus (muskrat)	Anal gland		√	√	Cyclopentadecanone, 5 *cis*-cyclopentadecenone, Cyclopentadecynone, Cyclohexadecanone, cycloheptadecanone, 5 *cis*-cycloheptadecenone, 7 *cis*-cycloheptadecenone, 5 *cis*, 11	3, 4, 5, 6		van Dorp *et al.*, 1973

Table 2.3 Continued

Genus species	Type of gland	Seasonal cycle of activity	Presence in sexes ♂ ♀	Chemical composition of secretion	Determination technique	Behavioural significance	Reference
1	2	3	4	5	6	7	8
				cis-cycloheptadecadienone, 5-cycloheptadecynone, 7 *cis*-cycloheptadecen-5-ynone, cyclononadecanone, plus all associated macrocyclic · alcohols			
Desmana moschata (desman)	Ventral			Saturated and unsaturated aliphatic and alycyclic ketones, with odd number of C atoms. (Main component saturated C_9-C_{19}, monoenic C_{17}-C_{23} and C_{21} dienic *n*-propylketone, as well as C_{15} and C_{17} dienic macrocyclic ketone.)	2, 3, 4, 5		Zinkevitch *et al.*, 1973

Notes

Column 2 Information listed is type of gland or secretion and anatomical position.
Column 3 √ = yes, × = no, ? = not certain, no entry = nothing known.
Column 4 √ = yes, × = no, no entry = nothing known. A circle around a tick, (√), denotes the sex in which the gland is best developed.

Column 5 Many different analytical methods and conditions have been employed in the chemical analyses (see Column 6), hence many studies are of qualitative value only. Cholesterol appears to be present in all mammalian sebaceous gland secretions and thus is omitted from mention in the table.

Column 6 1 = elemental analysis.
2 = thin layer chromatography.
3 = gas liquid chromatography.
4 = mass spectroscopy.
5 = infra-red spectroscopy.
6 = nuclear magnetic resonance analysis.

Column 7 An asterisk denotes those behaviours which have been *demonstrated*, rather than assumed, to have been influenced by the named chemical compounds.

Mammals apparently utilize a huge variety of substances, but fish utilize only proteinaceous substances and amino acids. Bacteria associated with the scent glands undoubtedly do contribute breakdown products (Gorman *et al.*,1974; Albone and Perry, 1976), but the role these substances play in intraspecific communication is suspect. It must be remembered that relatively few odorous secretions are simply thrown to the wind by the producer (though this is the usual broadcasting method employed by insects). Secretions are normally applied to a substratum, a rock, a piece of wood, another organism of the same species, etc. Not all the chemical complexity of a secretion is odorous; neither is it all behaviourally active (Hesterman, Goodrich and Mykytowycz, 1976). After the scent is set, there is a physical and chemical interaction between both the odorant and the non-odorant fractions and between the odorant and the substratum. Four factors appear to influence the volatility of an odorous secretion. These are: (1) the physical and chemical quality of the substratum; (2) the physical and chemical nature of the non-volatile constituents of the secretion; (3) the polarity of the odorant; and (4) the humidity of the atmosphere surrounding the site of odour deposition. In summarizing the resulting interplay, Regnier and Goodwin (1977) say:

> In general, non-polar odorants will show little interaction with polar materials (both *substratum* and *sebum*) and their evaporation will not be greatly influenced by humidity. On the other hand, polar odorants will interact with polar *substrates* and sebum and their volatility will be substantially reduced. Competition of polar odorants and water vapor for the same polar sites causes enhanced evaporation at high relative humidity. These results appear to correlate quite well with current explanations of chromatographic processes.

2.3 Threshold levels of perception

As with any perceptual system, the better the receptors the more sensitive to low-level stimulation will be the entire system. Broadly speaking, the vertebrates are macrosmatic, although the birds as a group are microsmatic. These terms refer not only to the amount of use to which the nose is put, but also the size of the nose and the area covered by olfactory membrane. The terms are misleading because they give no impression of specialized sensitivity to one or a group of odorants to which a structurally small system may have evolved. Not all mammals have highly structured noses, however, and the Orders Cetacea, Primates, Tubulidentata, Monotremata and Carnivora pinnipedia have relatively undeveloped olfactory apparatus. Man regards himself as microsmatic on the basis of these criteria (a view which finds support in his anthropocentric disdain for his reliance on olfactory cues), but studies on the sensitivity of his membrane show it to be every bit as good as the cat – a species acclaimed for its macrosmatic qualities. The difference in overall olfactory ability between the species is due to two factors. Firstly, the cat possesses an absolutely greater area of olfactory membrane than man (20.8 cm^2 and 11.5 cm^2) and hence more receptors, and, secondly, the membrane is far better situated to receive the inhaled odour-laden air streams (Fig. 2.6). But once the odorant molecules reach the membrane, only eight are required to trigger a single receptor in man; macrosmatic species can do

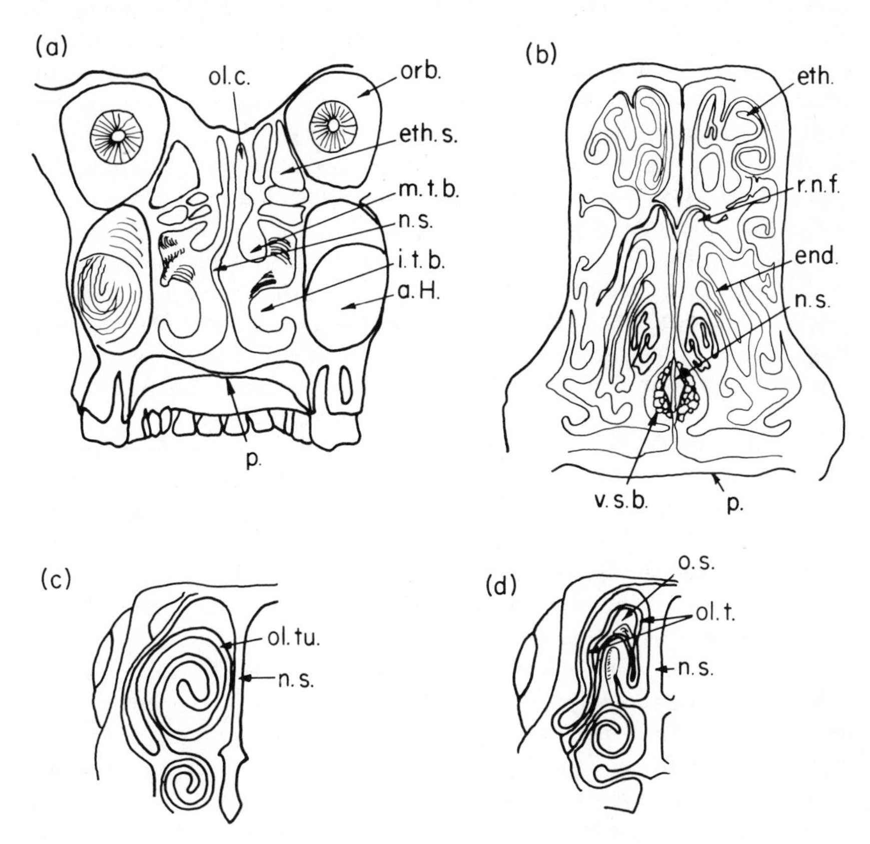

Figure 2.6
(a) Thick section through nasal fossa of man, viewed from behind. (Modified from Gray, 1905.)
(b) Section through snout of cat, anterior to orbit. (Modified from Negus, 1958.)
(c) Section through head of turkey vulture, *Cathartes aura*. (Redrawn from Portmann, 1961.)
(d) Section through head of fulmar petrel, *Fulmarus glacialis*. (Redrawn from Portmann, 1961.) a.H. antrum of Highmore; end. endoturbinals coated with olfactory mucosa and housed in nasal fossa; eth. ethmoturbinals coated with olfactory mucosa and housed in frontal sinus; eth. s. ethmoidal sinus; i.t.b. inferior turbinated bone; m.t.b. middle turbinated bone; n.s. nasal septum; o.s. orbital sinus; ol.c. olfactory cleft; ol.t. olfactory turbinal; ol.tu. olfactory tubercle; orb. orbit; p. hard palate; r.n.f. roof of nasal fossa; v.s.b. vascular swell body.

little better (De Vries and Stuiver, 1961). It is for these anatomical reasons that humans have a low power of acuity, but by any measure they have high powers of discrimination (i.e., the ability to tell apart one odour from another). Apart from the question of the ability of the brain to register, remember and interpret the odorous signals it receives, and there is some evidence that humans have quite well-developed abilities in this field, there are two main aspects to discrimination. One refers to the number of different chemical substances capable of resolution by the perceiving animal and the other concerns the ability of the perceiver to detect concentration changes of different odorants. Wright, in his lively account of the physics and chemistry of olfaction (1964), analyses the olfactory response of dog and man to a variety of odorants. He concludes that dogs perceive substances like acetic and butyric acid at concentrations a million times lower than man requires for perception.

With substances like mercaptan and ionone the dog's minimum perception level is only about a thousand times lower than man's perception level. He concludes that dogs are equipped to smell something which is denied to man with respect to the carboxylic acids, but are equipped to smell something similar to that which man can perceive with respect to the pungent, searing substances found in skunk and other mustelid anal gland secretions. Tests with rats under conditions of conditioned suppression indicate that humans can resolve smaller odour concentration changes than can rats (Davis, 1973). Pigeons, on the other hand, respond only to massive changes in odour concentration. On this basis the human, and presumably primate, olfactory system is superior to that of other mammals. As has been pointed out already, humans are able to locate the origin of an odour source by utilizing minute differences in odour concentration between one nostril and the other (von Békésey, 1964).

All this brings us to the question of sensitivity of the receptor units. There is little doubt that great differences in absolute thresholds of perception exist with respect to different groups and species of vertebrates. Fish have very low thresholds to many substances, primarily amino acids and related substances (Hara, 1976; Hara and MacDonald, 1976). The white catfish, *Ictalurus catus*, has a lower limit to perception of L-glutamine of between 10^{-7} and 10^{-8} M. Interestingly the D isomer of the same substance is far less effective in evoking a response (Suzuki and Tucker, 1977). Caribbean fish of the genus *Hepsitia stipes* and *Bathystoma rimator* respond to concentrations of creatinine, lactic acid and glutamic acid of as low as 10^{-9} M. These substances are normally present in the food of these fish (Steven, 1959) and so this ability is presumably used in the detection of prey. Eels have powers of perception of substances like β-phenyl alcohol or α-ionone of 3×10^{-18} M and 3×10^{-14} M respectively. This is no better than the ability shown by dogs for these substances, but neither fish nor mammal may be specifically adapted to their perception because they are the odours of roses and violets respectively (Teichmann, 1959). Dogs can perceive odorants at these low concentrations with ease. Solutions of 2×10^{-14} M acetic acid, 0.32×10^{-14} M α-ionone, 0.4×10^{-8} M caproic acid, and 0.5×10^{-8} M caprylic invoke perception at a minimal level (Wright, 1964). In each case, man requires concentrations four of five orders of magnitude higher. Rats can detect 2×10^{-13} M amyl acetate in air while humans require 2.8×10^{-10} M (Davis, 1973). Vampire bats (*Desmodus*) can detect between 3.9×10^{-3} M and 7.5×10^{-3} M butyric acid; the human threshold for this substance is 3.1×10^{-2} M (Schmidt and Greenhall, 1971). Vampires are rather special with regard to their olfactory acuity, for most European bats have thresholds between 10^1 and 10^7 times higher for substances like aliphatic aldehydes, acids, alcohols and ammonia. In general, the thresholds for neotropical species is lower than for temperate species (Obst and Schmidt, 1976). In a later chapter the phenomenon of olfactorily guided migration will be discussed and while there are absolutely no data available on the levels of odorant concentrations present in cues picked up by the homing animal, it can be assumed that they are very low indeed.

Is there any evidence that the threshold of perception changes as the individual animal grows older? For non-human species there is little, but for humans there is evidence of quite marked changes (Strauss, 1970). Just one substance was used as a test odour, phenyl ethyl alcohol, and this was chosen because of its known lack of effect on the trigeminal nerve and because it is an unfamiliar odour to most people. The alcohol was put in a modified wash bottle, the outlet pipe of which led to a double nosepiece. The subject was asked to hold his or her breath and a known small volume of air was injected into the bottle. If no odour was perceived the experimenters introduced a further one or two cc. of air into the bottle and this continued until the subject reported sensing the odour. The results of this experiment, using 100 children aged 8–10 years, 100 adolescents aged 16–18 years, and 100 adults aged 21–39 years, are shown in Fig. 2.7. It is at once clear from this trial that children and adolescents have far lower thresholds for perception of the odorant than adults, and the difference is highly significant at the 1% level. Children and adolescents also have far less variability about their means than adults indicating that some adults retain ability to perceive low threshold concentrations better than others. The interest in these results emerges when one considers them with what is known about changes occurring with age in the other sensory modalities. In vision and hearing the decrement in acuity is very gradual and shows no sudden leaps. Vision starts deteriorating in childhood and hearing at the start of the third decade. In addition the body's motor ability starts its decline during the third decade (Wenzel, 1948). What the factors are which are responsible for the decline in olfactory acuity remain unknown. The relationship between the nasal organs and both reproductive state and behaviour in animals is well known (Chapter 4), but as far as humans are concerned Freud has written that civilization has brought about a repression of the olfactory sense, at least in its involvement with reproduction. (Freud's collected papers, Vol. IV, London 1949). Alas, we know

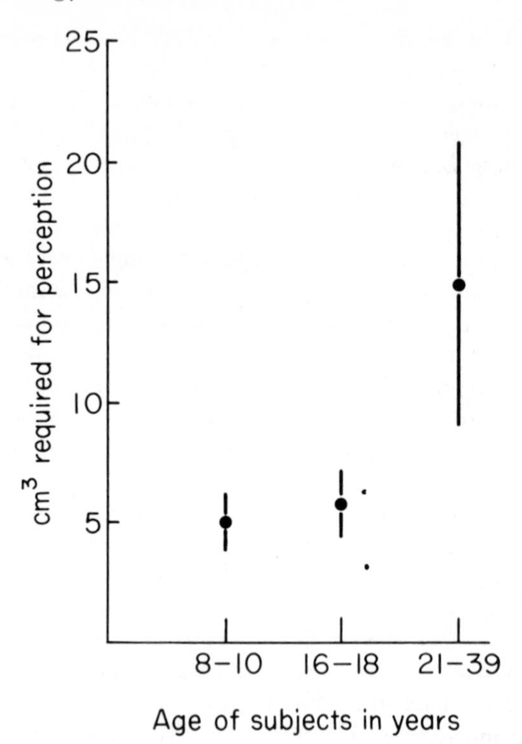

Figure 2.7 Sensitivity of human subjects of various ages to phenyl ethyl alcohol vapour. There were 100 subjects in each group and the group mean is computed from the individual medians of six repeated trials. Vertical line represents 1 S.D. The difference between the means for the two younger groups is not significant, but the difference between both the younger group means and the adult mean is significant, $p < 0.001$. (Adapted from Strauss, 1970.)

nothing of the changing olfactory requirements placed upon an individual as it develops through adolescence to sexual maturity, nor do we know whether its sensitivity to certain odours changes.

What is the relationship between receptor sensitivity and discriminatory ability? It is generally believed by physiologists that a receptor system optimized for the detection of lowest amplitude inputs is incapable of great feats of discrimination (Davis, 1973). Conversely, a system capable of high resolving power cannot detect very low amplitude input signals. The human eye contains two receptor unit types. The rods are poorly sensitive to wavelength inputs but highly sensitive to the lowest levels of illumination. The cones, on the other hand, have a high wavelength resolving power but low sensitivity to low levels of illumination. Humans may only have olfactory receptors analogous to 'cones' while rats may have only 'rods'. We do not know. Neither do we know whether both have the same receptor but quite different central processing arrangements.

3

Detection of food

One of the major components of the ecological niche concerns the food available to a species and the method that the species employs to exploit it. Few species, other than endoparasites and the parasitic males of some abyssal fish, such as the angler *Linophryne*, are surrounded by their food so that there is no need for them to orientate towards it, although all vertebrate embryos are in that happy situation. All other species have to be able to recognize their food species and distinguish them from others that may be less desirable, distasteful or even poisonous. The feeding interrelationships in community structure and integrity, which are its binding and distinguishing characteristics, are governed by this ability. While it is undoubtedly true that some species (e.g., many species of birds) rely almost exclusively on one sense for food detection, recognition and selection, the huge majority rely upon information received by all their senses. Olfaction plays a dominating role in the food-seeking behaviour and feeding ecology of nocturnal, crepuscular blind, burrowing, and cave-dwelling species with good vision, whether they be specialist or generalist feeders.

The olfactory system is called into service to secure food almost immediately following birth in mammals. In marsupials, birth occurs after a gestation as short as 12 days, and although the eyes and ears of the neonate are virtually functionless, the olfactory system is surprisingly well developed (McCrady, 1938). How the youngster finds its way from the cloaca to the pouch and onto a nipple has not yet been unequivocally demonstrated, but it would appear almost certain that olfactory cues have a lot to do with it. Later in life the same individual may live the life of a herbivore, perhaps relying rather lightly on odour cues, or as a generalist omnivore relying heavily on food odours. Young placentals are born usually at a more advanced state of development than marsupials but, even so, careful research has revealed that olfactory perception exerts a powerful influence on feeding behaviour during the first couple of weeks of life. If the ventral parts of a she-cat newly delivered of a litter are bathed and washed frequently, the newborn kittens experience great difficulty in finding a nipple. Cessation of washing restores their ability (Ewer, 1961). Ablation of the olfactory bulbs of young kittens renders them altogether incapable of finding a nipple (Kovach and Kling, 1967), but once their eyes are open orientation to the nipple presents no problems, even to bulbectomized subjects.

3.1 Innate response to food odours

The ability to follow certain odour cues is clearly developed at a very early age. Does that mean that the response to olfactory food signals is innate? The available evidence suggests that in some vertebrates the response to food odours is innate; in others it is not. But it must be stressed that a thorough analysis of the general principle has yet to be undertaken. Most young carnivorous mammals have to be introduced to their prey by their parents, and many complex behaviour patterns are necessary to achieve this (Ewer, 1968). In one series of experiments aimed at elucidating the role of olfactory sign stimuli in prey selection by polecats, *Mustela putorius*, twenty newly weaned young were divided into four groups and all were fed a diet of dead chicks (Apfelbach, 1973). One group was additionally fed dead rats and mice from the second month, and a second group was fed dead rats and mice from the fifth month. The third group received dead rats and mice from the eighth month, and the fourth (control) group was kept on dead chicks only. The result of this experiment was that only the first group readily ate the rats and mice. Those that were first introduced to this food after their fourth month paid decreasing attention to it. At one year of age, the control group polecats showed no response to the odour of rats and mice, though they continued to respond readily to that of chicks. Under natural conditions, it can be argued that during the first four months of life young polecats would have been introduced to the complete range of commonly occurring prey by their parents, and thus have learned the odorous sign stimuli associated with each prey item. The lack of feeding plasticity in the ecology of this species might account for its recent rarity, bordering upon extinction, in many parts of the British Isles. Serious habitat perturbation may have a deleterious effect on a species through that species' own lack of ability to seek out and exploit new prey.

The role of parents is not as well-developed in all species of vertebrate predator as it is in polecats. While adult snakes provide an overall service of protection for their young, they do not appear to assist their young very much in finding food. Much work has been performed on the response to odours emanating from potential prey shown by newborn or newly hatched snakes, and it offers an interesting comparison with that conducted on predatory mammals. Garter snakes of the genus *Thamnophis* respond to extracts of the surfaces of normally eaten prey by rapid tongue flicking followed by an attack. The number of tongue flicks following introduction of the extract and the latency until an attack is delivered can be measured with accuracy. Working with newborn eastern plains garter snakes, *T. radix*, Burghardt showed that annelids, both earthworms and leeches, provided the strongest stimuli followed by fish, larval salamander and frog. The same extracts presented to western smooth green snakes, *Opheodrys vernalis*, elicited quite a different pattern of response, with the extracts of slugs and crickets eliciting the strongest responses (Fig. 3.1) (Burghardt, 1967). These findings on naïve, inexperienced individuals coincide closely with what is known of the feeding ecology of these species of snakes in their natural environment, and indicate that the predisposition to certain species of prey is genetically determined. Further confirmation of this point comes from a consideration of the odour preferences shown by newborn snakes of one species which has a widespread distribution. One such is *Thamnophis sirtalis*; in fact this species of garter snake has the greatest geographical range of any North American ophidian. In Florida, it inhabits marshy and everglade habitats, and adults, i.e., experienced feeders, from this area respond avidly to extracts of fish (smelt,

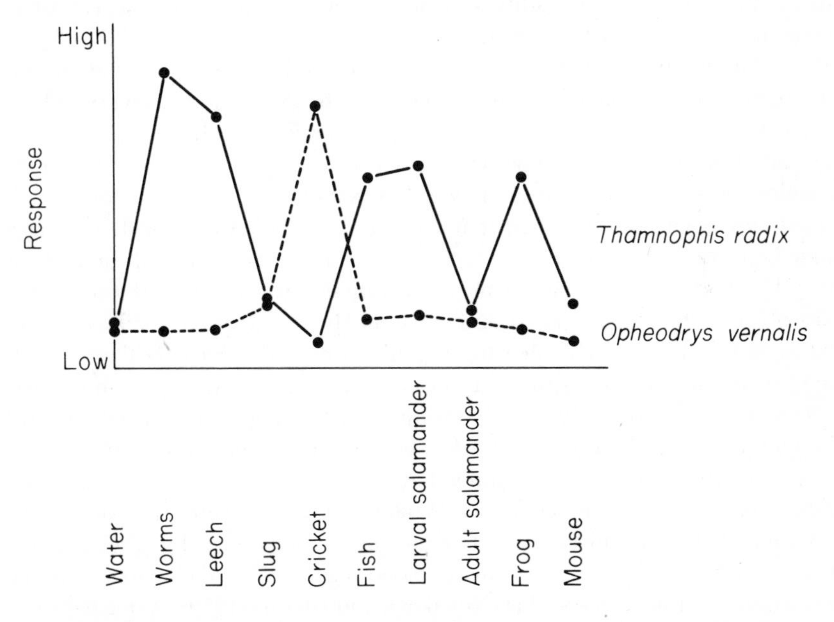

Figure 3.1 The response of two species of snakes, as measured by the rate of tongue flicking, to odours of various food species. Worms: *Lumbricus terrestris*, *L. rubellus*, *Eisenia foetida*; Leech: *Placobdella parasitica*; Slug: *Deroceras gracile*; Cricket: *Acheta domestica*; Fish: *Notropis atherinoides*, *Lebistes reticulatus*, *Carassius auratus*; Salamander: *Ambystoma jeffersonium*; Frog: *Acris crepitans*; Mouse: *Mus musculus*. (Modified from Burghardt, 1967.)

Osmerus mordax); so too do newborn snakes. Further north, in Massachusetts, adults show scant interest in fish extracts, here their habitat is much drier and lacks large bodies of standing water. Newborn northern garters likewise express a low preference for fish odours. The different odour preferences shown by the two geographical morphs are innate and indicate genetic differences within the one species (Dix, 1968). There are indications that lizards possess an innate ability to discriminate between the odours of suitable and unsuitable foods. Newly hatched southeastern five-lined skinks, *Eumeces inexpectatus*, show a strong rejection of extract of assassin bugs, but readily orientate towards extract of click beetle larvae and extract of mouse. The rate of tongue flicking is interesting, for the highest rates are obtained with reference to assassin bugs and the lowest for click beetles and mice; the odour brought into the vomeronasal organ on the tongue would thus seem to assist the lizard in its choice of a suitable meal. Assassin bugs are highly predatory on other insects and may be mildly poisonous to reptiles (Loop and Scoville, 1972).

This is not to say that the feeding preferences of all reptiles are shackled to their genetic makeup and ecological niche. The newborn of many species respond eagerly to odours of food items not normally eaten. The young of Butler's garter snake, *Thamnophis butleri*, readily respond to odours of fish and happily eat fish in captivity, yet fish never figure in the natural diet of this species (Burghardt, 1967). Observations like these indicate that variability of response to odour does occur in newborn snakes of some species which, given sudden severe habitat perturbation, might prove to be of selective and ecological advantage.

The diet chosen by a species is one which satisfies all its nutritional requirements. It has long been known that rats utilize their powers of smell to select between a vitamin-deficient and a vitamin-complete diet. Such a choice is based on experience, the rat having learned the characteristics of the fulsome diet. Destruction of the olfactory nerve of a dog results in aphagia; the dog chews any object or food and rejects it indiscriminately (Le Magnen, 1959, *inter alia*). No such rejection occurs when the trigeminal or glossopharyngeal nerves are destroyed, thus indicating that it is olfaction, rather than taste, which controls acceptance of food. Olfactory sensitivity is strongly influenced by the individual's state of hunger, and it has been found in humans that the level of olfactory acuity is at its highest immediately preceding a meal and at its lowest right after a meal (Hammer, 1951). Whether olfactory acuity is related to the production of gastric juices or regulated by them is not known, but a relationship between insulin production, and hence blood sugar level, and olfactory stimuli is known to occur in dogs and rats (Nováková and Dlouhá, 1960). A dog deprived of food and food odours for several hours exhibits a constant blood sugar level. The odour of finely chopped meat brings about a sudden fall in blood sugar level caused by a release of insulin. Olfactory bulbectomy in rats results in the development of a diabetes-like disease, though how the exclusion of olfactory signals to the brain actually brings about the disruption in insulin supplies is not clear.

3.2 The detection of plant food by odorous cues

Herbivorous animals have neither to pursue nor attack their food, and probably it is for this reason that little attention has been paid to the mechanisms whereby the choice of the particular species, or part of an individual plant, is made. This lack of attention is curious because of the known effect exerted by grazers and browsers on grassland and bushland seral ecology. The forage sought by herbivores is discrete by virtue of dispersion and time of ripening so, although it appears that they have to do no more than simply drop their heads to find food, in fact, they have to actively seek it out, much as carnivores do. Some of the most remarkable known examples of vertebrates actively seeking out plant products are seen in the bats. South American fruit bats of the species *Carollia perspicillata* feed only upon ripe bananas and will seek out their food from within human dwellings and even by night. These fruits are consumed where they grow or are stored, but the leaf-nosed bats of the genus *Artibeus* specialize in carrying off the fleshy nuts of *Acromia* trees for later consumption. They only take the ripe fruit which they appear to detect by its smell, since no visible change is apparent. Many nuts are dropped on the journey and the kernels fall to the ground around the roosting sites; the bats play a role of some importance in the dispersion of this species of tree. Nectar-feeding bats of the family Glossophaginae also locate sources of ripe nectar by its odour, and in probing for it act as pollinators. Flowers sought out by bats tend to be drab in colour and quite different from the brightly coloured species which are sought after by birds such as hummingbirds and sun birds, and also have wide, shallow nectaries and a firmly built landing platform. Most significantly they have a strong odour (to the human nose) and bloom at night when birds do not fly (Porsch, 1932). Nectar feeding ceases abruptly if a wind should suddenly spring up during the night, and on a windy evening the arrival of bats at the nectar flowers is significantly delayed. On calm nights there is no disruption of feeding, even if the nights are rainy or foggy. Indeed, it is on such nights that *Pteropus* makes its most devastating mass attacks on fruit-bearing trees. Further evidence that these bats find their food by odour is derived from a consideration of their olfactory bulbs. The bulbs of the Glossophaginae are far better developed than in insectivorous bats (Allen, 1930; Møller, 1932), and on the basis of the relative volume of the brain concerned with olfaction they are 10% better endowed than insectivorous bats (Fig. 3.2) (Suthers 1970). An interesting parallel is afforded by the only nocturnal species of fruit-eating bird, the oilbird of northern South America, *Steatornis caripensis*. The similarity of this species of nightjar to bats is quite remarkable, for it utilizes a form of echolocation for navigation around its roost caves. It has long been recognized that this species has well-developed olfactory tubercles and bulbs (Bang, 1960). It seems highly probable that the trees bearing ripe aromatic and spicy fruits on which the oilbird feeds are located by smell, as trees bearing almost ripe fruits are overlooked (Snow, 1961, Stager, 1967). There are no other birds which, from a

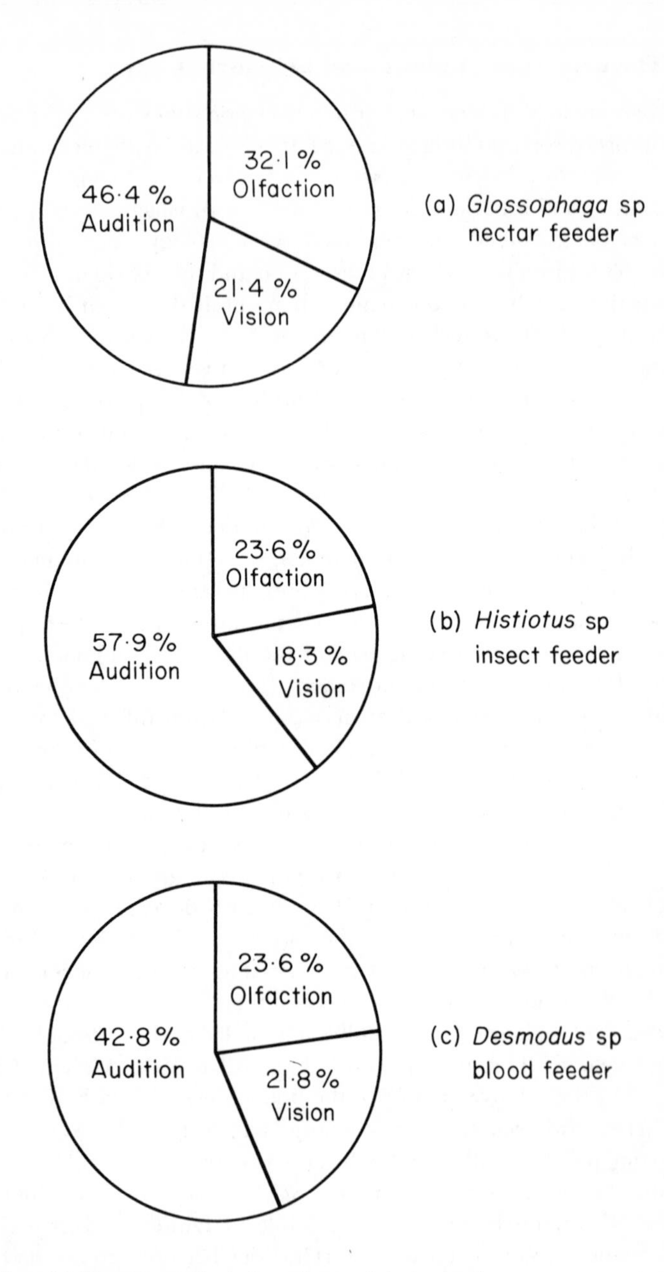

Figure 3.2 The estimated relative volumes of the brain of
(a) a nectar-feeding,
(b) an insect-feeding, and
(c) a blood-feeding bat concerned with audition, vision and olfaction.
(From Suthers, 1970; after Mann, 1961.)

study of their olfactory apparatus and ecology, are likely to use their noses in the location of plant food, with the exception of geese (*A. anser*) (Würdinger, 1979).

Many fishes commonly feed on plant foods at night and in turbid waters when visibility is restricted. Although no experimental work appears to have been done on plant food selection by fish, it has been shown that fish such as the blunt-nosed minnow, *Hyborhynchus notatus*, can discriminate between waters in which different plants have been rinsed. Blinded minnows respond just as well as sighted ones, but those whose olfactory epithelium has been cauterized and rendered functionless cannot effect the discrimination. Odour sensitivity to plants is extremely high and suggests that herbivorous fishes could use their powers of odour discrimination to find suitable food plants (Hasler, 1957).

A few attempts have been made to investigate the role of olfaction in plant food location by terrestrial mammals, and there are also many anecdotes which lack scientific investigation. For example, it is thought that mammals such as the arctic hare, *Lepus arcticus*, and the Norway lemming, *Lemmus lemmus*, which spend the winter underneath several feet of snow, find their food plants by their aromas. In these conditions of total darkness vision is impossible (Bourlière, 1955). Small forest-dwelling rodents feed heavily upon fallen seeds and nuts of trees. Frequently, and usually late in the winter or early spring, these food items become covered by leaf litter and buried under the soil surface by the action of rain and frost. Small rodents possess a refined ability to locate such hidden food stocks. Sunflower seeds and hazel nuts can be dug up from as deep as 30 cm by the mouse *Apodemus flavicollis*. If the nuts are ground up into a flour prior to burial, the rodents locate them and dig for them as if they had been whole – clear evidence that tactile cues play no part in location behaviour (Sveredenko, 1954). This ability presents real problems to foresters trying to reseed cleared land. Depending on the time of year and the size of rodent populations, between 70–100% of newly sown conifer seeds can be removed by rodents (Howard, Marsh and Cole, 1968). The palatability of conifer seeds is positively correlated with seed weight; the large seeds of Douglas fir, *Pseudotsuga taxifolia*, being more attractive than the small seeds of sugar pine, *Pinus lambertiana*. Treatment with odorous mineral oil, Lecithin, has no repellent effect. Rodents respond to the odour of foods when no food is actually present. In an elegant series of experiments conducted some years ago it was found that ground squirrels, *Spermophilus beecheyi*, spent 15 min in a 15-h test period near to an orifice from which came the odour of oats, 2.5 mins near a wheat source and 0.8 mins near a source of odourless air (Howard, Palmateer and Marsh, 1969). In the wild, oats is a preferred food of *Spermophilus*.

There is some evidence, then, that plants are located by specific odours emanating from them. Perhaps future research will show whether the herbivorous marine iguanas from Galapagos, *Amblyrhynchus cristatus*, select seaweed species by odour, and whether the many herbivorous mammals exercise food selection by using odour cues. It is easy to imagine how the leaf-eating arboreal marsupials select their tree species by odour, for the distinctive aromas of these eucalyptus species are readily detectable by our own noses. It is less easy to imagine how the small, selectively feeding antelopes make their choice of grass species, for these plants have little obvious odour, at least to the human nose. Detection of hypogean fungi by dogs and by forest-dwelling vertebrates almost certainly depends upon olfaction. Some species, e.g., the bettong, *Bettongia lesueur*, rely heavily upon hypogean fungi during the period of the year when other fruits and berries are unavailable.

Species which scatter-hoard their food, e.g., many squirrels, the acouchi, *Myoprocta pratti*, etc., apparently relocate it using odorous cues. Frequently the food is marked with scent by the burying animal and the odour of this mark prevents others from digging it up. Glands situated at the corner of the mouth (the *angulus oris* glands) are thought to be the source of this special scent tag (Muul, 1970). Observations of different family groups of acouchis indicate that the mark applied to a food item by one member of a group is recognized, and the food consequently left alone, by other members of the same group. If the mark is applied by a member of a different group, the food item is quickly dug up, remarked, and reburied. The selective advantage of this behaviour under natural conditions is obvious.

Browsing and grazing ruminants exhibit a precise form of food plant selection, often selecting species of high nutritious value. When presented with a range of unfamiliar food plants, deer exercise their choice by first closely smelling the branches. Once the goodness of a particular food species has been tested in this way, it is subsequently recognized by sight and no further sniffing of the branches occurs (Longhurst, *et al.*, 1968; Jackson, 1974). It is unlikely that deer can actually smell the nutrients contained within the forage and much more likely that they respond to volatile 'indicator' compounds which are associated with desirable nutrients. Many plant volatiles are inhibitory to the correct functioning of the rumen microflora and it follows that the most aromatic ought to be most inhibitory to rumen function. Experimentation has shown this to be so and the most unpalatable plant species are the most inhibitory. Since the discernible chemistry of a plant relates closely to that species' defence ability, what the ruminant is doing in not consuming a species unattractive on account of the odour of its volatiles is to assess and interpret its defences.

3.3 The detection of animal food by odour cues

If the paucity of experimental evidence leaves some doubt about the breadth of involvement of olfaction in the detection of plant food, there is no doubt about its role in the detection of animal food. Live animal prey may be pursued by a predator following an invisible, but odorous, trail, and stalking species are careful to approach a potential quarry from downwind. Since prey detection, and predator avoidance, are the product and result of odour production and perception, it would be advantageous to a prey population to exhibit odour polymorphism. Certainly, human beings have individual odours, this being the basis upon which tracker dogs work, and different families and populations of rodents have demonstrably different odours (Stoddart, Aplin and Wood, 1975; Godfrey, 1958). Visually hunting predators tend to concentrate upon the common colour varieties of prey and to overlook rarer forms even if they are very obvious. This is because of the development of a specific search image; a behavioural quirk enabling a predator to hunt swiftly for its prey, the characteristics of which it has learned. Such behaviour is responsible for apostatic selection which maintains colour polymorphism within the population. There is no evidence to argue against the maintenance of scent polymorphism within populations and some evidence for it. Mice can be trained to feed on dough pellets tainted with either peppermint or vanilla, and when so trained they retain this induced preference when offered both. In other words, the training period allows them to assume a specific olfactory search image which they can hold before them when confronted with a polymorphic population (Soane and Clark, 1973). Apostatic selection would maintain the polymorphism in subsequent generations.

Many insectivorous species of vertebrate feed on prey which is concealed beneath the surface of the soil. Amphibians and reptiles are, on the whole, exceptions to this, although there are a fair number of burrowing insectivorous snakes. Sufficient is known about feeding in frogs and toads to indicate that anurans use a combination of vision and olfaction for food finding. Toads are able to find mealworms when relying only on olfaction, and when presented with objects smelling of mealworms they show the same sort of appetitive behaviour as they do when feeding on prey which is visible to them. Under these conditions they exhibit much displacement behaviour, 'yawning' and the like, because the visual stimuli normally emanating from the prey and releasing the final consummatry act are missing (Heusser, 1958; Sternthal, 1974; Hemmer and Schopp, 1975).

Insectivorous blind snakes, *Leptotyphlops dulcis* and *Typhlops pusillus*, are specialist feeders on termites and ants, but are not immune to the attacks of these creatures. Normally they follow the trail odours left by the ants and sometimes actually travel in the columns. They are thus more or less surrounded by their prey but are never molested. Yet a blind snake taken from a vivarium in the laboratory and introduced experimentally into a column is sorely set upon by the ants. As it coils and writhes it spreads all over its body a secretion produced by the cloacal glands which acts as a repellent to further ant attack. Blind snakes are usually pale brown dorsally and pinkish-white ventrally, but during ant attacks they turn somewhat silvery on both dorsum and ventrum. A similar colour change is seen in the African blind snakes, *L. scutifrons* (Visser, 1966). In *L. dulcis* this change in colour is brought about by a partial raising of the scales, having the presumed function of offering a roughened surface into which the protective cloacal gland secretion can be rubbed. After the period of intense writhing the original coloration reappears, but the whole body has a glossy appearance caused by the secretions (Watkins *et al*, 1969). Several other functions are served by the secretions, however. They not only protect the snake from attack by the ants, but they also protect the snake from its major enemies, ophiophagous snakes, such as *Lampropeltis triangulum*, and also from the competitive interest of other insectivorous snakes such as *Sonora episcopa* and *Tantilla gracilis*. They are attractive to other blind snakes of the same species and some even exhibit courtship behaviour when exposed to it. Blind snakes are positively attracted to the odour emanating from the capital glands on the workers of, e.g., *Neivamyrmex nigrescens*. They respond neutrally to skatole which is released when workers are mutilated; in this respect blind snakes differ from all potential competitors which are strongly repelled by the smell of skatole. The ants' capital gland secretion has the dual function of attracting other ants and repelling certain insectivorous snakes. Army ants use trail pheromones to coordinate their activities and these substances repel certain predators. These substances are attractive to blind snakes who follow them to their source. This complicated ecological interaction is summarized diagramatically in Fig. 3.3.

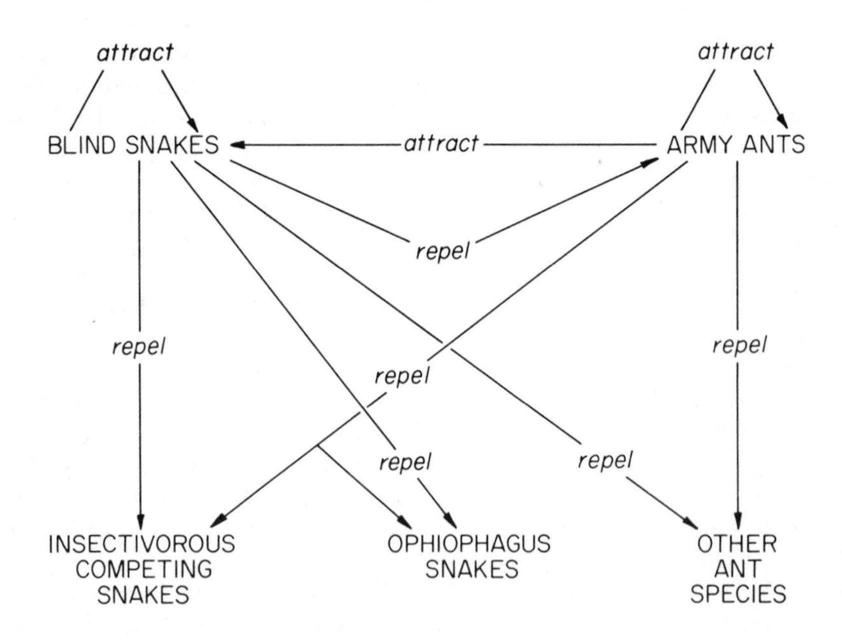

Figure 3.3 Diagram of olfactory interactions in the ecological interrelationships between blind snakes, *Leptotyphlops dulcis*, insectivorous competitors of blind snakes, their predators, army ants, and other ant species. (Modified from Watkins, Gehlback and Kroll, 1969.)

The majority of insectivorous birds locate their prey by vision and by tactile sensations, and the pressure receptors in the bills of probing species are highly developed for this purpose. Two species of birds, however, locate their invertebrate prey by olfaction. The kiwi, *Apteryx australis*, is a probing, nocturnal insectivorous bird with limited powers of sight and the structure of its olfactory system has already been mentioned. In one experiment conducted in a large aviary, five- to six-inch-long aluminium tubes were sunk into the ground with their rims flush with the soil surface. They all contained soil but some also contained pieces of chopped earthworm mixed with the soil. A screen of butter muslin fixed over the top of the tube prevented the kiwis from tasting the soil. An inspection of the tubes and their muslin caps each morning showed that only the tubes containing the earthworms had been probed, the controls being left untouched (Wenzel, 1968). The only other insectivorous birds for which olfaction is important are the honey guides, *Indicator indicator* and *I. minor*, the olfactory tubercles of which are very large and fill most of the third respiratory chamber. Friedmann quotes the diaries of an early Portugese missionary in East Africa who noted that his church building became highly attractive to honey guides a short while after he had lit some beeswax candles. The birds flew right up to the candles and fed on the soft, warm wax (Friedmann, 1955). More recent studies have confirmed these observations made over three centuries ago. (Stager, 1967). A beeswax candle placed in a tree attracted no honey guides while it remained unlit. But within 35 minutes of lighting it had attracted half a dozen birds to feed on the warm wax. The flame was screened from sight.

Considering the insectivorous mammals as a group, it seems that heavy reliance is placed on olfaction for the location of food. Although insectivorous bats utilize their specialized hearing for pinpointing the position and course of airborne insects, some olfactory discrimination of palatability occurs. The olfactory part of the brain of insectivorous bats is smaller than in fruit bats, but it nevertheless accounts for 24% of its volume (Fig. 3.2). Many small bats, e.g., *M. myotis*, *M. emarginatus*, *Eptesicus serotinus*, *Nyctalus noctula*, *Plecotus auritus* and *Pipistrellus nathusii* are able to locate hidden insects by smell from a few centimetres range and to reject unpalatable species. Even when beetles are suspended from a thread, *M. myotis* is able to distinguish palatable from unpalatable species by their odours when it is in flight. An edible beetle hung ten centimetres away from an unpalatable beetle is avoided, so pervasive is the odour of the distasteful insect (Kolb, 1958, 1961). Terrestrial insectivores have excellent noses and rely heavily upon them to lead them to suitable food. In one interesting study on the ability of common shrews and North American deermice (*Peromyscus* sp.) to find buried sawfly cocoons, it was clear that the small mammals could not only tell exactly where a cocoon was buried and almost invariably dig for it in exactly the correct place (Fig. 3.4), but they could also tell what sex of occupant the cocoon housed. The cocoons of female sawflies are larger than cocoons of males, and under field conditions many more female

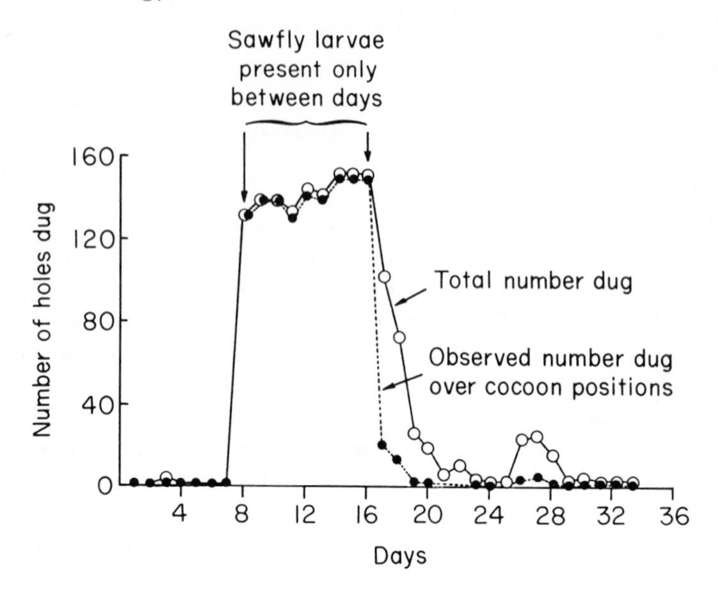

Figure 3.4 The number of holes dug by deermice and shrews for buried sawfly pupae. (Redrawn from Holling, 1958.)

cocoons are attacked than males. The odour cues emanating from these structures must contain a great deal of information which is decodeable by a foraging insectivore (Holling, 1958).

From the very beginnings of recorded history it has been assumed that vertebrate meat-eaters, i.e., predatory carnivores, use their noses for locating their prey. This assumption is certainly correct, although most species combine olfactory with visual, and other input to make an effective attack. The relative importance of the various types of stimulating input differs quite widely between classes of vertebrates, and even between members of the same class. Dogfish, *Scyliorhinus caniculus*, have long been thought to rely exclusively upon their highly sensitive olfactory systems for the detection of potential prey items (Fig. 1.8*k* and p. 27), but more recent studies on the same species have revealed that dead fish meat can be located purely by its odour, while live fish are located in a different way. In experiments designed to reveal what this was, the prey was placed inside a low flat box which was buried in the sand on the bottom of an aquarium containing a hunting dogfish. A current of water was forced over the prey in the box and, when the box contained only dead fish, the dogfish repeatedly attacked the outlet stream. When the box contained a live plaice, the dogfish attacked the spot under which the box lay and ignored the outlet. It would appear as if a dogfish can locate live prey by detecting the action potential of its muscles (Kalmijn, 1971). Sharks are, nevertheless, highly sensitive to dissolved food odours and presumably utilize this ability to help them find injured prey. If pieces of fish flesh are held in a concealed perforated plastic tube and introduced into a water current, sharks rapidly appear from downstream and gather excitedly at the site of odour emergence. Sharks can follow an 'odour corridor' very precisely; in one set of experiments utilizing whitetip sharks, *Triaenodon obesus*, an injured but mobile parrot fish was allowed to swim along a tank. After it had reached the end of the tank, and cover, the shark was released and it was able to follow exactly the jinking movements of the parrot fish (Hobson, 1963, Tester, 1963a). Much attention has been paid to the feeding behaviour of sharks for obvious reasons relating to human survival in shark-infested waters, and the question arises whether sharks can detect and locate intact undamaged prey by olfactory cues alone. Blinded blacktip sharks, *Carcharhinus melanopterus*, respond to a cage containing an undamaged grouper fish by eliciting a normal feeding attack; no such attack is elicted by an empty cage. If water in which live, undamaged fish have been kept is siphoned silently into a tank containing a blinded blacktip shark, the shark responds rapidly with a typical hunting reaction. Such a shark is oblivious to a siphon stream of control water. Thus it seems as if sharks can detect undamaged prey by some substance, or substances, given off from the body surface. The physiological condition of the prey also appears to be important. Sharks which circle excitedly around a siphon spout bringing water from a concealed vessel containing fish quickly lose interest if the fish remain quiescent. But they show intense renewed excitement if the fish are stressed by being harassed by the experimenter. Shark excitement mounts in relation to the degree of stress elicited in the prey and reaches a maximum after the prey has died. Thus it seems possible that sharks can detect healthy prey by its odour, but they respond more eagerly to prey which is physiologically stressed to some extent (Tester, 1963b).

Odour corridors are important for all species of fish which locate their prey by olfaction. Undisturbed water and diffusion fields of odour are most unlikely to occur. In practice, fish use the current as the orienting stimulus and the current can be created by the movement of the prey itself. This is an essential aspect of host detection by lampreys and bullheads, *Ictalurus natalis*. In this latter species, severance of the olfactory nerve does not result in total inability to find food, but the typical to-and-fro movements become slower and punctuated by much overshooting, as if the fish is unable to keep in the corridor.

Food finding by smell is common amongst amphibians. Newts of the genus *Triturus* have been observed to follow closely a trail made by an earthworm. A diagram of a bout of trailing behaviour is shown in Fig. 3.5, and it can clearly be seen that when the newt strays away from the trail it quickly swings back to refind it (Herter, 1941). Natterjack toads, *Bufo calamita*, have been shown to be able to find mealworms entirely by smell, and when presented with materials smelling of mealworms show the same sort of appetitive behaviour as they do towards real mealworms. However, toads apparently require a visual stimulus for the consummatory act and if this is lacking show much displacement activity

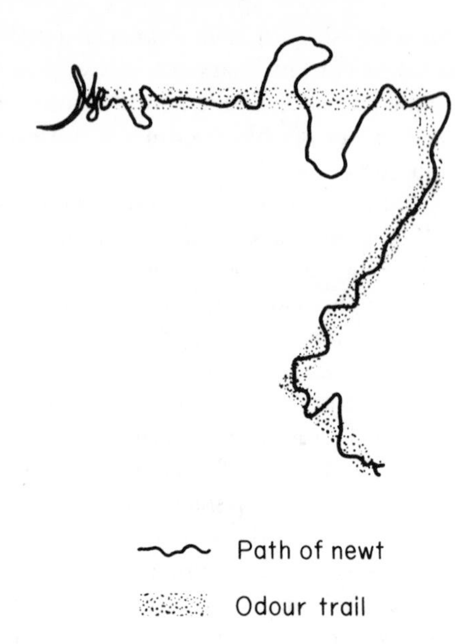

Figure 3.5 Plan of the trail of a newt, *Triturus* sp., following an odour path made on the floor of an arena by drawing along an earthworm. (Redrawn from Herter, 1941.)

such as 'yawning' and snapping at non-existing objects (Heusser, 1958). Natterjack toads display an increased responsiveness to prey if they can smell it; scant notice is paid if the prey item can only be seen and not smelled (Hammer and Schopp, 1975). Further evidence of the role of olfaction comes from experiments on leopard frogs, *Rana pipiens*, which, when presented with mealworms rendered noxious by being dipped in dilute hydrochloric acid (1.85% v/v) and rosewater, exhibit a significant delay in attack response by the third day. It appears that the first two days of the experiment serve as a conditioning period during which the frogs learn to associate the smell of the rosewater with the unpleasant taste of the mealworm (Sternthal, 1974).

There is an abundance of evidence illustrating just how important the sense of smell is to hunting snakes and a point of great significance emerging from many studies is that the vomeronasal organ of Jacobson is deeply involved. Removal of Jacobson's organ (by cauterization) or removal of the tongue (by surgery) of a grass snake, *Natrix natrix*, results in a total disruption of trailing behaviour. An intact snake that is allowed to 'mouth' a mouse that was subsequently killed by the experimenter and dragged across an arena quickly discovers the trail and follows it closely to the kill at the end of it. A mutilated snake searches round at random and at times passes very close to the dead mouse without recognizing it and totally fails in its task (Kahmann, 1932). In other species there is differential effect wrought by mutilating either the tongue of Jacobson's organ. An early and elegant series of quantified experiments using brown snakes, *Storeria dekayi*, and garter snakes indicated that cutting off the tongue tips and covering the nostrils resulted in a total loss of detection of food. Covering the nostrils only resulted in a four-fold delay in detection compared to the behaviour of a normal, intact snake. Cutting off the tips of the tongue only delayed detection twofold (Noble and Clausen, 1936). The details of these experiments are shown in Fig. 3.6.

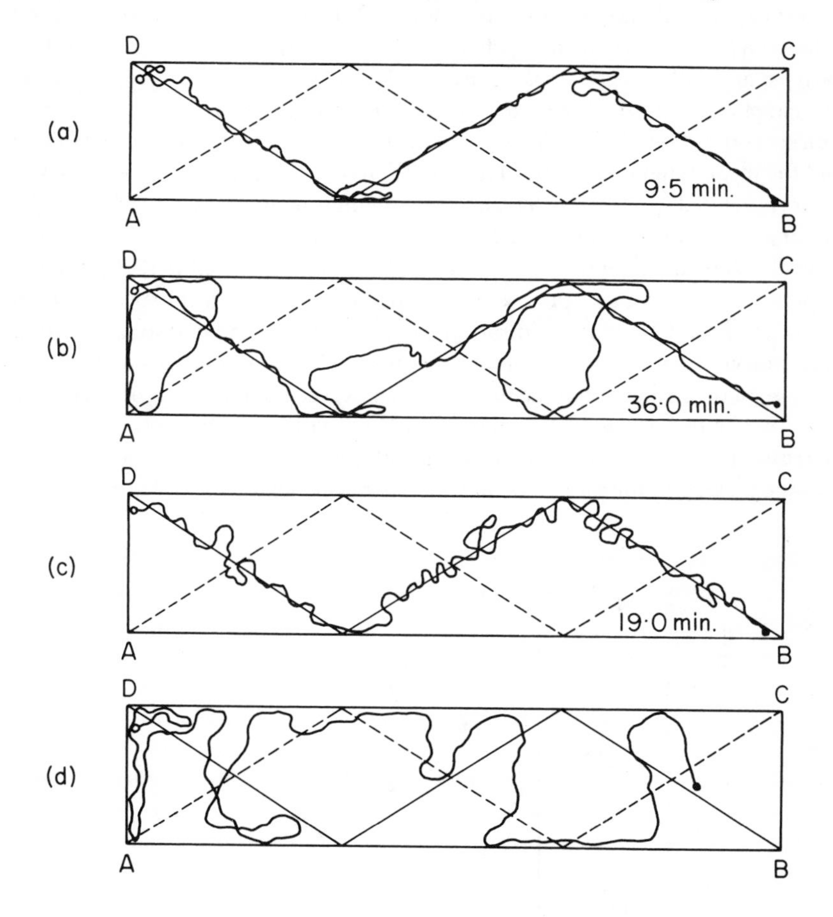

Figure 3.6 Diagram of soapstone surfaces. Continuous line in the area, the food-scented trail; broken line, the distilled water trail.
(a) The trail of normal *S. dekayi* beginning at D and ending at B, with the time required to cover the trail;
(b) the trail of an *S. dekayi* with olfaction eliminated;
(c) the trail of an *S. dekayi* with Jacobson's organ incapacitated;
(d) the trail of an *S. dekayi* with both Jacobson's organs and the olfactory organs incapacitated. No time reaction is listed here, since the trail was not followed from D to B and the time involved here was not significant. (From Noble and Clausen, 1936.)

The precise role of the bifid tongue is far from clear. If the entire tips are cut from the stump, contact by the stump with the body of a prey specimen or a swab bearing its body odour still elicits an attack response. Complete removal of the stump does not preclude attack behaviour if the lips of the snake are allowed to make contact with the prey or surrogate prey (Wilde, 1938). Nevertheless, the tongue is extensively used by hunting snakes in part of their approach-identification-striking complex of behaviours. It is also much involved in trailing behaviour, the behaviour in which a trail left by a prey item is followed. When garter snakes, *Thamnophis sirtalis*, are timed for their ability to navigate a maze at the end of which is a food reward – a piece of earthworm – it is found that there is a direct correlation ($r = -0.85$) between mean running time for individual snakes and the mean tongue-flick rate (Fig. 3.7) (Kubie and Halpern, 1975). Snakes of a closely related species, *T. radix*, exhibit a higher tongue-flick rate when in the proximity of odour of newt or fish, and the difference in rate between this and a control is statistically significant (Chiszar, Scudder and Knight, 1976). Such a response to the odour of prey is, not surprisingly, absent from species which detect their food by sensing infra-red radiation. In a similar testing situation rattlesnakes, *Crotalus viridis*, and massasangas, *Sistrurus catenatus*, showed no increase in their rates of tongue flicking when exposed to the odour of mice. The sight of a mouse, however, whether with or without mouse odour greatly increased the rate of tongue flicking in both newborn and adult rattlesnakes. Tongue flicking in the pit vipers may be more associated with striking and prey

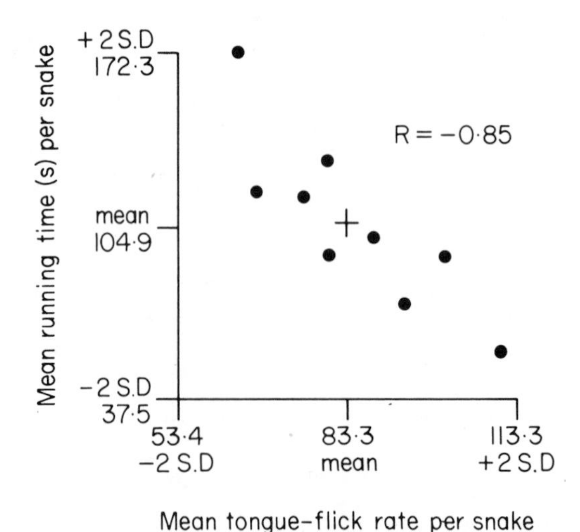

Figure 3.7 The relationship between the time taken to traverse a maze towards a food odour by individual garter snakes and their rate of tongue flicking. (From Kubie and Halpern, 1975.)

swallowing than with prey detection. Pit vipers, like so many venomous snakes, strike the prey and immediately let it go. The envenomated prey runs for a few seconds before collapsing, whereupon the snake trails it. European vipers, *Vipera berus*, when offered mice they have themselves killed and others killed by an experimenter, show a selective preference for the envenomated mice. It is thought the preference is olfactorily based and stems from substances produced by the effect of the haemolytic and cytolytic venom on the prey's tissues (Baumann, 1927, 1928). Perhaps it is at this stage in the feeding behaviour of pit vipers that the organ of Jacobson is used.

There are few known examples of predatory birds using olfaction for the detection of their prey, but it appears that the predatory ability of several, and perhaps all, of the Procellariformes (petrels and shearwaters) depends upon their powers of olfaction. When a rag soaked in cod liver oil and another soaked in water are positioned on floats at sea, Wilson's and Leach's petrels approach the oil about 16 times during 30 minutes of observation and never once approach the control rag. Under captive conditions, when presented with sponges soaked in the same two substances, Wilson's petrel and the greater shearwater show a statistically significant difference in their choice of the oil-filled sponge. Leach's petrel does not show so marked a choice difference and the sooty shearwater chooses the water sponge twice as often as the oil sponge (Grubb, 1972). Since many of the procellariform birds forage at night, vision is unlikely to be of major importance; their demonstrated olfactory ability would seem to have a high adaptive significance.

Quite surprisingly, little work appears to have been performed on the reliance placed by predatory mammals upon olfaction, but there can be little doubt that olfactory cues are of great importance, particularly to nocturnal predators. Observations on hunting behaviour of large cats, for example, show that the wind is continually sniffed and that the predator approaches from downwind. This serves to give the cat an odour gradient against which to work as well as to keep its own odour clear of the prey. Wolves, *Canis lupus*, sometimes find their prey by picking up odour trails, but they are just as likely to use vision to spot either the prey itself or its fresh tracks in snow (Mech, 1970). Spotted hyaenas, *Crocuta crocuta*, on the other hand, never approach prey, or carcasses for that matter, from downwind, so it appears that in this species odour cues are not utilized in food detection. Social encounters, however, are always initiated by one individual approaching another from downwind, indicating that odour detection plays a role in this behaviour (Kruuk, 1972). Quantitative and experimental studies on carnivorous mammals are rare. The food selection abilities of polecats referred to on p. 64 indicates that this species is perfectly capable of hunting by scent alone, once it has been conditioned, when young, to the odour of a particular prey species (Apfelbach, 1973). Most predators practice only the mildest form of specialization and doubtless utilize all possible forms of sensory input in the detection of their prey.

3.4 Scavengers

Most flesh-eating species scavenge on the kills of others for some of the time; there are very few obligate scavengers and just as few obligate predators. Amongst the mammals it is thought that only the cheetah is a true predator, never utilizing the discarded kills of other species. But even this species returns to scavenge upon its own kills (Ewer, 1974). Amongst the birds the vultures represent a family of true scavengers, and their dependence upon olfaction has been debated since the time of Pliny and Aristotle. In fact there are two distinct evolutionary lines of vulture, quite distinct from one another. The cathartine vultures occupy the New World and the aegyptine vultures the Old World (Gurney, 1972). Old World vultures apparently rely little upon olfaction, but there have been no controlled experiments performed. Controlled experiments have only been performed upon the New World turkey vulture, *Cathartes aura*, and they show that this scavenger has a well-developed olfactory sense. A stuffed deer placed in the prairie fails to attract the attention of vultures, but a partially decomposing specimen of the same species and in the same attitude quickly draws the attention of the birds. That their interest is won by odours and not visual cues is illustrated

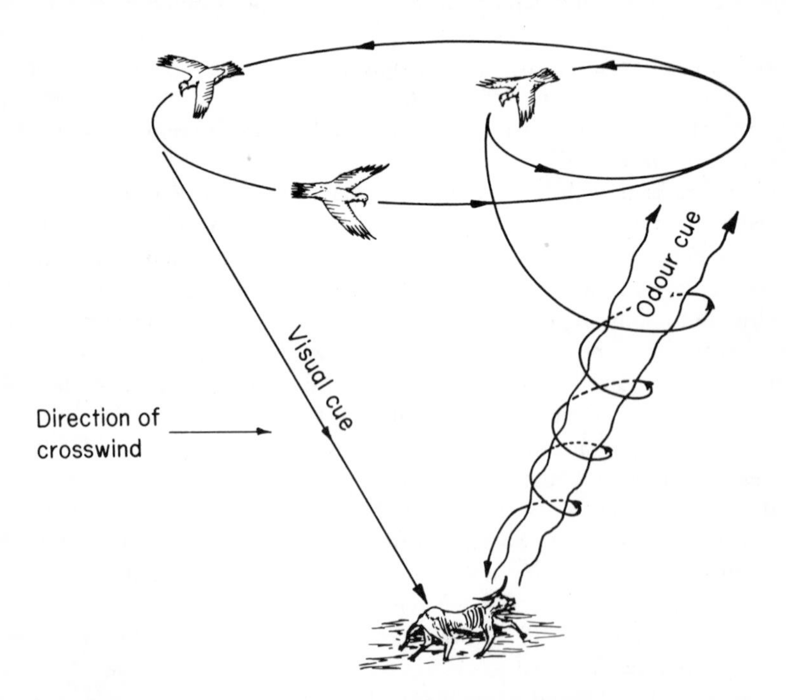

Figure 3.8 Food-finding flight path of the turkey vulture, *Cathartes aura*. Visual cues are used to alert the bird to the presence of potential food; odours emanating from it confirm its existence, restrict the flight pattern, and the vulture homes in on the carcase from downwind. (After Stager, 1964.)

by their active response to a plume of air drawn over a vessel containing carrion (invisible to the vultures) and released into the atmosphere. The vultures circle the resulting inverted cone of odorous air even if it has been substantially displaced by crosswinds (Fig. 3.8). Other species, notably *Coragyps*, descend and inspect on visual cue only. Comparative studies of other vultures indicate that only the central American king vulture, *Sarcoramphus*, utilizes olfaction in food detection; the rest of the New World, and apparently all of Old World, vultures find food only by vision and are unaffected by highly odorous, but hidden prey (Stager, 1964).

Red foxes, *Vulpes vulpes,* are broad-spectrum feeders which function as insectivores, predators and scavengers. Quite often foxes bury food remains in a small scrape made in the soil at the base of a tree. After burial it is normal for the fox to urinate on the ground, either directly over the spot or immediately adjacent to it. In field studies on wild foxes the urine marks serve as complex odour beacons to provide the fox with some information about the hidden food morsel upon its return. The classical interpretation of this scenting behaviour is that the odour guides the returning fox to its food cache, but recent detailed studies suggest the explanation is not so simple. Foxes especially mark inedible remains, such as bones, feathers and the like, and the time they spend investigating such fruitless caches is significantly less than the time spent in investigating caches from which only food odour cues emanate. Urine odour seems to indicate that it is a waste of time to attempt to unearth what lies below and so serves to enhance scavenging efficiency. Interestingly enough the urine marks of a different individual are just as effective as a fox's own marks and thus play a part in maintaining the feeding efficiency of all the foxes inhabiting the area (Henry, 1977).

3.5 Quasi-parasites

The vertebrate way of life does not lend itself to a parasitic modification and, if the bizarre example of adult male abyssal fish, parasitic (in the true sense of the word) upon their mates is ignored, there are only a few instances of vertebrates even partially exploiting that particular ecological niche. The lampreys and hagfishes have gone quite far in utilizing the tissues of living fish, and experimental studies with the lamprey, *Petromyzon marinus*, show that victim detection is made possible by orientation to its body odour. Just one or two amines seem to be the cue to which the lampreys respond. The data shown in Table 3.1 indicate that not only is the lamprey's response to trout body odour innate, but that, when rendered anosmic, lampreys show no response to the body odour of fish. The response of lampreys to fish is dependent upon the season and age of the parasite. From June to December in their first year young lampreys strongly seek out host fish, but as sexual maturity approaches in early spring and with it the spawning migration, lampreys feed little and do not respond to fish

Table 3.1 Response of the lamprey, *Petromyzon marinus*, to body odour of the trout. (Modified from Kleerekoper and Morgensen, 1963.)

Origin and previous history of test lamprey	Number of tests	Locomotor response to introduction of trout water to experimental tank	
		No response	*Response*
Normal lampreys			
(a) Taken from fish in Great Lakes	63	11	52*
(b) Reared in laboratory in total absence of fish	4	0	4
Anosmic lampreys			
(a) Taken from fish in Great Lakes	5	5	0

* $\chi^2 = 26.7$, $p = < 0.001$

body odour. The changing physiological state of the lamprey related to its state of sexual maturity influences the olfactory threshold for the specific chemical cues emanating from fish body secretions and excretions (Kleerekoper and Morgensen, 1963; Kleerekoper, 1969).

Another genus which closely approximates to the parasitic way of life is *Desmodus*, the vampire bats. The commonest species, *D. rotundus*, has a well-developed internal nasal structure having four endoturbinals, three ectoturbinals as well as naso and maxilloturbinal bones. These bats have a much lower detection threshold for butyric acid, which is an important constituent of mammalian sweat, than man. They react to levels as low as 3.9×10^{-3} vol % and could easily detect their large mammal prey at night by utilizing olfactory cues (Schmidt and Greenhall, 1971). Vampires do urinate copiously on their victims, however, and the possibility exists that they relocate their victims by following the mixed scents of their own urine and butyric acid. Nothing is known about how they are able to relocate the precise wound made on a previous feeding sortie; it was thought they could respond to the odour of dried blood, but this has now been shown to be ineffective; cattle experimentally treated with oxblood are no more nor less attacked than untreated controls (Turner, 1975).

Before concluding these remarks on the implication of odours in food detection by vertebrates, it is relevant to examine briefly the effect of the food ingested on the quality of the odour produced by the ingesting animal. Whether any change in the body odour of a herbivorous species induced by a dietary modification in any way affects the attractiveness of that species to its own predators is not known, although it is unlikely that it does. What it does do is to influence intraspecific attractiveness. Guinea pigs, *Cavia porcellus*, discriminate between conspecifics on the basis of their previous dietary history. The urine from other individuals kept on a diet which differs from that of a test guinea pig is investigated for a significantly longer period than urine from individuals fed the same diet (Beauchamp, 1976). This may be generally applicable to rodents, for it has also been shown to occur in Mongolian gerbils, *Meriones unguiculatus*, and rats which both belong to a distantly related rodent group. In gerbils, the secretions from the ventral sebaceous glands also are changed by dietary changes, and these differences are detectable by test subjects. In rats the quality of the maternal pheromone (see Chapter 4), which is produced in the caecum, is directly influenced by the caecal contents, and young rats will respond to the caecum contents of their mothers more readily than the contents of another lactating female kept on a different diet (Leon, 1975; Skeen and Thiessen, 1977). These observations indicate that the selection of food may play a more fundamental role in some aspects of social integration than has hitherto been thought.

3.6 Summary

Trophic relationships characterize community structure and integrity, and the ability shown by vertebrates to recognize, assess and seek out certain favoured food species maintains these relationships. In many species, the response to food odours is innate, although some exhibit a critical period in their early development during which they learn to recognize the odours of certain food species. Severe environmental disturbance may cause rarity or even extinction of formerly abundant species through the changed spectrum of food species, to the odour cues from which adults may not be able to adapt. In some species with wide distribution ranges, which occupy slightly different ecological niches throughout their range, there is evidence in naïve juveniles of an innate predisposition towards the odours of those prey species observed to be favoured by their parents.

The problems faced by herbivores in selecting food are essentially the same as those faced by carnivores, for although plant food does not have to be pursued and attacked, it has to be detected and its ripeness or other quality assessed. There is evidence from all classes of vertebrates of the role played by olfaction in allowing a correct selection to be made. The ability is particularly well developed in the fruit- and nectar-eating bats and in those rodents which feed on buried seeds. The only bird which uses its olfactory powers to locate plant food is the oilbird. Scatter-hoarding mammalian species relocate food caches by following odour cues emanating from scent marks placed on the food items by the individual hiding the food.

The majority of small vertebrate carnivores utilize olfaction in the location of food, though the only bird known to seek in this way is the kiwi. Insectivorous snakes secrete specialized substances, the odours of which attract their insect prey (army ants) but repel both competing insectivorous snakes and ophiophagous snakes. While attracting their prey, the substances also protect the snakes from attack by the ants. Many bats which normally feed upon aerial prey can locate insects which are hidden underneath litter on the ground using odour cues diffusing outwards from them. Sharks are able to discriminate not only the presence of prey fish, by detecting dissolved substances in the water, but also the physiological state of the prey.

Scavengers rely heavily upon odour cues for the location of hidden carcasses; visual or acoustic cues are not produced from such food sources. Turkey vultures do not respond to the sight of a dead animal unless and until they perceive the smell of decomposition. Foxes mark hidden carrion by urinating, but the odours serve to prevent themselves and other foxes from wasting time by searching for inedible remains.

Lampreys certainly, and vampire bats possibly, do use olfaction in the selection of a suitable host. Although vampires show a high sensitivity for butyric acid, a component of mammalian sweat, they always urinate copiously on the host and relocation of the host may be effected by a complex odour analysis.

The incidence of olfaction in food detection and selection is seen to be widespread throughout the class Vertebrata and is not restricted just to those species living in darkened environments. Some species invariably find their food by following odour cues, such as lampreys, snakes, kiwis and shrews, while others initially utilize odour cues but subsequently learn to recognize palatable species by sight, such as deer, rats and possibly many herbivores. Since the food preference of each species remains roughly the same from one generation to the next, and so maintains the integrity of the ecological niche of that species, it follows that recognition of the characteristics of the food must either be innate or be taught to the newborn young early in its life. An innate knowledge of food characters is seen in many species lacking intense parental care and which do not pass through a feeding transition from a juvenile to an adult diet, e.g., snakes, while in those species which have well-developed parental care and do not start to feed upon the definitive diet until after being weaned off a juvenile diet, the characteristics of its food is learned from the repeated presentation by the parents of *their* choice of food. The importance of a thorough understanding of the feeding ecology of animals is clearly apparent when it is necessary to make an assessment of the power and resilience of natural ecosystems, an exercise carried out more frequently now than formerly as a result of the ever-increasing pressures made on the environment. The power an ecosystem possesses to repair itself and its resilience in the face of attacks by disease organisms or human-induced imbalance depends upon the structure of its internal trophic relationships, which ultimately depend upon the precise food choices made by its component species.

4

Reproductive processes

The forces of natural selection which act upon the reproductive products of any generation are not always the same as those which acted upon earlier generations, for changing ecological conditions and the occurrence of random mutation create a different substratum upon which they can act. The resulting acquisition of new features, both physical and behavioural, is the keystone to speciation. There is much evidence illustrating the role that olfactory perception plays in maintaining species integrity and this is examined in Chapter 5, but in this chapter the influence of odour in the various component stages of breeding, as a population dynamic character, and reproduction, as a physiological character, is examined. Rather more is known about the involvement of olfaction in the reproductive biology of mammals than in any other class of vertebrate; this results, without doubt, from the massive amount of research directed toward the mammals in the pursuit of knowledge about reproductive physiology and behaviour and, indirectly, from the application of wide-spectrum animal husbandry procedures. Isolated examples from the other vertebrate classes, with the exception of the birds, suggest that olfaction is equally heavily committed in breeding and reproduction. It is clear, therefore, that an understanding of the importance of olfaction in this area is a necessary, and perhaps crucial, part of the design of pest control and animal husbandry studies.

The very first events in the reproductive overture must be advertisement of sexual status and mutual attraction of the sexes. Without the close physical juxtaposition engendered by attraction, breeding cannot occur. Attraction usually starts when one of the sexes comes into breeding condition, either through age-related developmental maturation or through environmentally influenced recrudescence of the reproductive glands. As might be expected from consideration of a large assembly of species with differently developed sense organs, attraction is expressed through visual, auditory and olfactory cues. In most species, more than one and possibly all the senses are involved, although it is obvious that deaf species, like many snakes, or blind species, like cave-dwelling salamanders and some burrowing mammals, cannot use sensory systems they do not possess. Provision of a working back-up sensory system ensures that reproductive processes do not cease should accident or injury deprive an individual of sight, hearing or smell. This makes experimentation very difficult, for when a test animal is subjected to selective sensory deprivation procedures, the behaviour may still be observed to occur, stimulatory input coming through an alternative sensory channel.

Attractive and advertising signals in vertebrates are not under the sole control of reproductive chemistry. Many examples are known of pelage and plumage changes which are socially induced associated with the attainment of high social status (e.g., coat colour in the blackbuck, *Antilope cervicapra*, and plumage in the Fulmar petrel, *Fulmarus glacialis*). As will be seen in Chapter 6, odours change in relation to social, as well as to sexual, status. The complexity of the controlling mechanisms makes model building extremely difficult, but the widespread occurrence of odorous switches and rheostats within the sub-phylum nevertheless indicates a reproductive control mechanism of great evolutionary antiquity.

The occurrence, seasonality and presence in one or both sexes of a variety of vertebrate scent glands is shown in Table 2.3. In very few species are glands present only in females; in the majority they are found only, or are better developed, in males. Very many scent glands only develop at the time of sexual activity and are sensitive to male sex hormones. By way of illustrating this point, Fig. 4.1 shows the activity and size of the dorsal gland of the North American kangaroo mouse, *Dipodomys merriami*. Although the gland is present in both sexes, it is nearly vestigial in females. From its winter resting size in males it increases almost five-fold during the breeding season (Quay, 1953). Similar increases in size are seen in the chin glands of non-migratory amphibians such as *Plethodon*. During winter these glands are poorly developed and quite small (Madison, 1977). Snakes and lizards, while well-endowed with scent glands, do not appear to exhibit a marked increase in gland size, although the submandibular glands of crocodiles press outwards through the skin during the breeding season and secrete copious amounts of pungent musk (Ditmars, 1910; Guggisberg, 1972); the same is true of the tortoise, *Gopherus berlandieri*, which also bears submandibular glands (Rose, 1970). In both these reptiles the glands develop only in the breeding season. At least one worker on birds has suggested that the uropygidial gland of some species functions as an odour-producing organ (Paris, 1914), although the suggestion has not been tested experimentally. It is of some coincidence that it is perhaps best developed in the kiwi (*Apteryx australis*), a

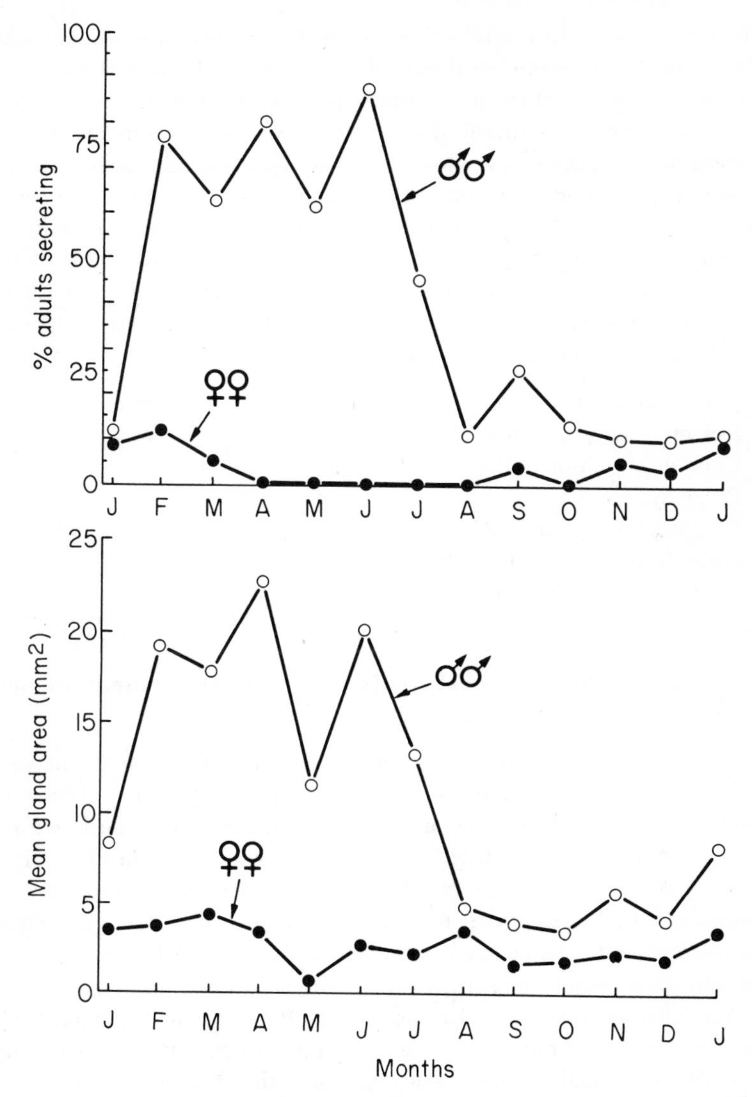

Figure 4.1 The relationship between dorsal gland size, activity and time of year in *Dipodomys merriami*. (Modified from Quay, 1953.)

species renowned for its remarkable olfactory capability (see Chapter 3, page 73).

Anatomical evidence such as this suggests that olfaction may be implicated and involved in the reproductive processes of a wide range of vertebrates, for many, indeed most, of which there is as yet no behavioural evidence for its involvement. It is too early yet in the development of our concept of odour control mechanisms in reproductive processes to say whether there is a single basic model, the specific features of which have been modified by the environmental and ecological conditions imposed upon the species, or whether there is a multiphyletic origin of olfactory control and manipulation of these processes. The adoption of a parsimonious approach would favour the former alternative and a good argument could be made for it. This chapter is devoted to a critical appraisal of the six main aspects of reproductive biology in which olfaction plays a role. Species recognition and the establishment of breeding systems are aspects of the reproductive process which are excluded here on account of their breadth. They are dealt with in Chapters 5 and 6 respectively.

This chapter covers the following topics:

4.1 Sex attraction and recognition; the advertisement of sexual status.
4.2 Detection and induction of oestrus, ovulation and lordosis.
4.3 Courtship, mating and related behaviours.
4.4 Pregnancy.
4.5 Parental behaviour; imprinting.
4.6 Growth, physical and psychosexual development.
4.7 Summary and conclusions.

4.1 Sex attraction and recognition; the advertisement of sexual status

There cannot be a single owner of an unspeyed female cat or dog who needs to be convinced that it is the odour of their charges during the time of heat which induces all the male cats and dogs in the neighbourhood to pay court. Attractive-smelling substances are carried in the urine and bitches urinate far more frequently when on heat than when they are pregnant or in anoestrus (Beach and Gilmore, 1949). Male dogs spend significantly more time investigating urine from oestral females than urine from anoestral females and they react differently (Graf and Meyer-Holzapfel, 1974).

Table 4.1 shows the results of a study designed to examine the quality of bitch urine. Bitches are attracted to urine from male dogs, but their reactions are mostly limited to sniffing only. Male dogs sniff urine from bitches on heat only, and show no other reaction, on less than 10% of occasions, so clearly much information is contained within the urine. The adaptive significance of such a strong scent lure is clearly seen for these solitary-dwelling carnivorous species which defend territories. Because of the sometimes huge areas defended, the marks must bear their message for quite some time. It has been reported that the attractant, pungent substance found in the urine of female tigers retains its activity when sprayed onto vegetation and tree trunks for up to four weeks (Shaller, 1967). If the human nose can perceive it for that length of time, it is likely that the tiger can do at least as well and most probably better. It is of

Table 4.1 The behavioural reaction of male dogs to the urine of bitches in oestrus and anoestrus. (Modified from Graf and Meyer-Holzapfel, 1974.)

Reaction of adult male dog	Percentage of trials in which male dog's reaction is shown to urine from	
	Bitches in oestrus	*Bitches in anoestrus*
Urination on top of, or very close to, urine mark made by bitch	83.3	58.3
Scratching ground by urine mark made by bitch	58.3	8.3
Licking pool of urine made by bitch	66.6	8.3

interest to note, by the way, that the olfactory powers of the tiger have been known for a long time for the Sanskrit name for that species, *Vyagra*, is derived from a verb stock which means 'to smell' (Brahmachary, personal communication).

Fishes of the abyssal depths live at very low densities and have overcome the problem of mate attraction by the male being very small and parasitic upon the female, as was described in Chapter 1. As larvae, male ceratioid angler and other bathypelagic fish live in the surface waters of the tropics and subtropics. Looking much like other pelagic fish larvae, their eyes are well-developed and used for feeding and orientation. At metamorphosis, however, great changes occur. The olfactory lamellae and associated structures, as well as the olfactory bulbs and forebrain, greatly enlarge (Fig. 1.2). Their eyes soon start to atrophy as the young males sink lower into the deeps. Although the evidence is circumstantial and inferred from the anatomical structure of the males, it is likely that the males locate the females, who have but poorly developed olfactory organs, by an odour cue emanating from them. Since anything that moves is a potential meal to a female deep sea angler, it is not known how the tiny male is able to attach to the special saddle, or prominence, which lies between the eyes, without being eaten (Marshall, 1971).

More gregarious species of fish do not have the same problem in making contact with others of their own species. Frequently their problems are restricted to ascertaining the sex of other individuals and then to attracting them into the initial stages of courtship behaviour. Males of many species of fish are attracted to substances emanating from females of their own species; indeed, the fishermen of the Mississippi basin have long used caged female catfish, *Ictalurus punctatus*, to lure large numbers of males to their traps (Timms and Kleerekoper, 1972). Among game fish, male sea trout (*Salmo trutta*) react strongly to water which had previously held ripe females and a similar, though not quite such a strong, reaction is seen in rainbow trout (*Salmo gairdneri*) (Solomon, 1977; Newcombe and Hartman, 1973). There is little evidence of anything more than attraction to bring the sexes together in close enough proximity to allow courtship to proceed.

Female newts of the species *Taricha rivularis* and *Triturus vulgaris* produce substances from the cloaca which soon attract the attention of males downstream. Water in which males only have been kept fails to elicit this response, indicating that the presence of dissolved substances enables sex recognition to occur (Twitty, 1955). The newt, *Plethodon jordani*, occupies fixed home ranges of between 2 m and 14 m maximum length. During the breeding season females show a marked preference for the chin gland odours of non-neighbouring males than for neighbouring males. This selection may reduce inbreeding. Under the same experimental conditions during the non-breeding period of the year, no preference for non-neighbours is seen. Whether the odorous cue is chemically changed in the breeding season or whether the receiver characteristics are changed is not known; either way not only is the sex of another animal determined but its proximity of residence is also discriminated (Madison, 1975). There are no well-founded reports of odorous sex recognition in the Lactertilia, although it seems unlikely that odours play no part in precourtship activity. The snakes, however, rely very much more on odour cues, and recognition of sex through this medium probably occurs in all families. The odour from the dorsal and lateral body skin of female garter snakes, *Thamnophis sirtalis*, and female brown snakes, *Storeria dekayi*, induces trail-following behaviour in males. Odours from the same region of males or females not yet in breeding condition fail to elicit this behaviour in both species (Noble, 1937). In many species of snake, large numbers of males are attracted to a single female and give rise to a sexual aggregation or mating group, but such groups do not occur in aggressive species. The function of these mating groups is not known (Noble and Clausen, 1936; Bennion and Parker, 1976). In many, and possibly most, species of terrestrial chelonian, the sex of another individual is determined by the odour of the chin or lateral ('musk') glands. The lipid component of the chin gland secretion of male *Gopherus berlandieri* is attractive to females who approach closely a model daubed with chin gland extract; the same extract elicits a violent head-bobbing and ramming behaviour in males (Rose, 1970). Females are also attracted to the cloacal gland secretion; males show no response. The genus *Geochelone* lacks chin glands; male *G. carbonaria* and *G. denticulata* approach all large, tortoise-like objects with suspicion. If the object is another male, a head-bobbing challenge is met with a similar response and a ramming fight ensues. If it is a non-responding female, however, the challenging male approaches its rear end and sniffs its cloaca. Cloacal secretion from receptive females is sufficient to elicit courtship behaviour in males when spread on inanimate models (Auffenberg, 1965, 1969). In the musk turtles of the family Kinosternidae, the sex of an individual is determined by the males who approach others and sniff the tail and musk gland region. Males and non-receptive females move off; only receptive females stand fast to enable mounting to take place (Mahmoud, 1967).

When considering the mammals it is not easy to distinguish between sex discrimination and attraction as the behaviours immediately preceding courtship are often complex and subtle. They will be treated together. Much work has been performed on the pig and this species might be taken to be a generalized model for the mammals. Several parts of the model have been demonstrated for other (wild and domestic) species, but in no other species have studies of such thoroughness been conducted. Both male and female pigs are attracted to the whole body odour of adequate sex partners (i.e., sows in oestrus and intact, adult boars) far more than they are to the odours of sows in anoestral or castrated or prepubertal boars. The attraction remains as strong and is impaired neither by hiding the stimulus animal from view nor by anaesthetizing it. After having had their olfactory bulbs removed, females in oestrus cannot distinguish between a boar and a sow on the basis of sight and hearing alone. Females in anoestrus pay no more attention to the odour of an adequate sexual partner than to an inadequate partner. As has been mentioned on p. 36, both sexes of pig possess large salivary glands which, under the influence of sex steroids, secrete odoriferous substances. The final proof that attraction in the pig is chemical can be seen when a testosterone-treated sow in oestrus is highly attractive to another, but untreated, oestral sow. This suggests that the metabolism of androgens in the salivary glands of both male and female is similar; what differs under natural conditions is the presence and absence of androgens (Signoret, 1976). These findings are summarized as a general pictorial model in Fig. 4.2 and Table 4.2.

Table 4.2 Sex discrimination and attraction in the pig (after Signoret, 1976).

Subject	*Number of experiments*	Mean number of seconds during a 5-min test spent near: *Adequate sexual partner*	*Inadequate sexual partner*
		Intact male	*Castrated male*
Oestral sow: stimulus animals visible	16	192.25	41.75
stimulus animals invisible	13	159.54	38.62
stimulus animals invisible and anaesthetized	14	148.14	64.14
Anoestral sow: stimulus animals visible	44	81.86	88.27
		Androgenized female	*Control female*
Oestral sow: stimulus animals visible	109	173.44	38.01
		Oestrous female	*Anoestrous female*
Intact boar: stimulus animals visible	235	140.39	113.06

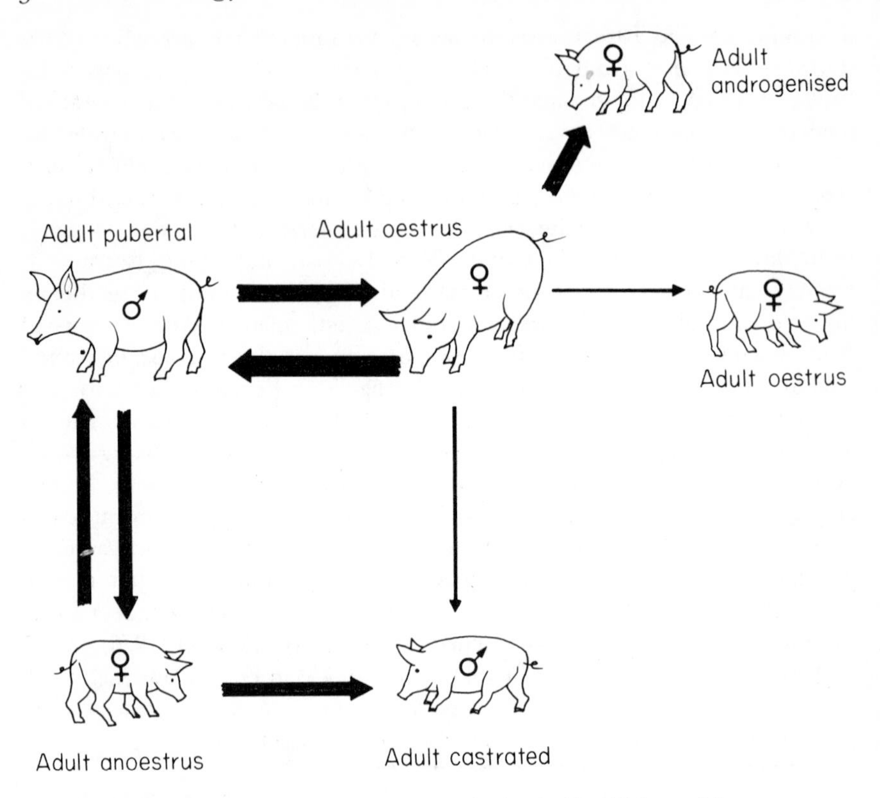

Figure 4.2 Model of odorous sexual attraction in the pig. The thickness of the arrows denotes relative strength of attraction. (Based on data in Signoret, 1976.)

Studies with sheep reveal that rams determine sexual condition of ewes, and hence sex itself, by perceiving the odour of the vaginal secretions. Rams make repeated sexual advances towards anoestral ewes if they are first smeared with vaginal secretions taken from oestral ewes. Rams from which the olfactory bulbs have been removed fail to discriminate between adequate and inadequate sexual partners. Ewes are equally attracted to the odour of rams and a smaller proportion of those rendered anosmic by removal of the olfactory bulbs are mated than of intact control ewes. Various studies with blindfolding and deafening rams reveal that visual and auditory cues in sheep play a role of substantial importance (Inkster, 1957; Lindsay and Robinson, 1961; Lindsay, 1965, 1966; Fletcher and Lindsay, 1968). Fig. 4.3 graphically summarizes these results.

As far as the small, gregarious mammals are concerned, vaginal secretions and, to a lesser extent, urine exert quite powerful attractions during the period of oestrus. There is no absolute unity of experimental interpretation, for in some species the past sexual experience of the male influences his attraction while in

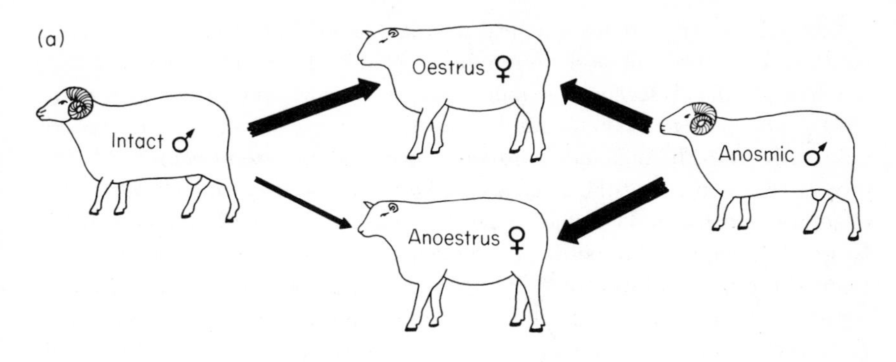

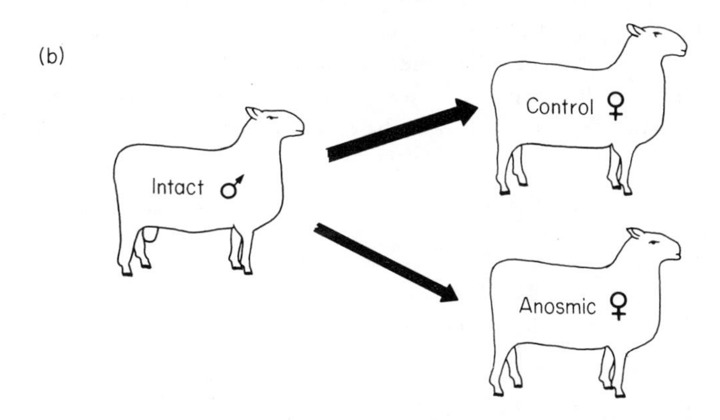

Figure 4.3 Model of odorous sexual attraction in the sheep.
(a) Number of sexual approaches made by anosmic and intact rams to oestral and anoestral ewes.
(b) Proportion of ewes mated by intact ram.
The thickness of the arrows denotes relative strength of attraction. (Based on data in Lindsay, 1965 and Fletcher and Lindsay, 1968.)

other species it does not. Sexually experienced male mice are drawn to the vaginal secretions of females in oestrus; sexually naïve male mice show no preference for secretions from oestral mice over anoestral or dioestral mice (Hayashi and Kimuri, 1974).

The same is true for rats (Carr, Loeb and Dissinger, 1965), but in this species the urine of females in oestrus also acts as a strong attractant to males (Lydell and Doty, 1972; Pfaff and Pfaffman, 1969). Although some attractive qualities have been ascribed to female mouse urine there is little doubt it is less effective than rat urine and much less effective than mouse vaginal secretion (Davies and Bellamy, 1972; Hayashi and Kimuri, 1974).

Vaginal secretion from an oestral female mouse daubed onto the perineal region of a dioestral female elicits mounting behaviour in experienced male mice but no such effect is seen with female mouse urine in use (Hayashi and Kimuri, 1974). It has been found in the rat that the pre-optic region of the brain responds differently from the olfactory bulb when urine odours are presented to a test subject. By monitoring the electrical spike discharge from single units in the olfactory bulb and in the tertiary olfactory neurones (Fig. 1.10), it is clear that a far greater differential response to non-urine odours occurs in the olfactory bulb than in the pre-optic region. Urine odours from oestral and ovariectomized female rats cause a statistically significant differential pre-optic response. Recordings from castrated male rats show the same trend (Pfaffman, 1971; Pfaff and Gregory, 1971).

What these studies reveal is that the differential olfactory response to sex odours occurs not in the primary olfactory neurones but higher 'upstream' in the system. That castrated males respond in a manner similar to that shown by intact males indicates that the hormonal state of the receiver is not a matter of fundamental importance. Fig. 4.4 indicates the extent to which the differential response occurs.

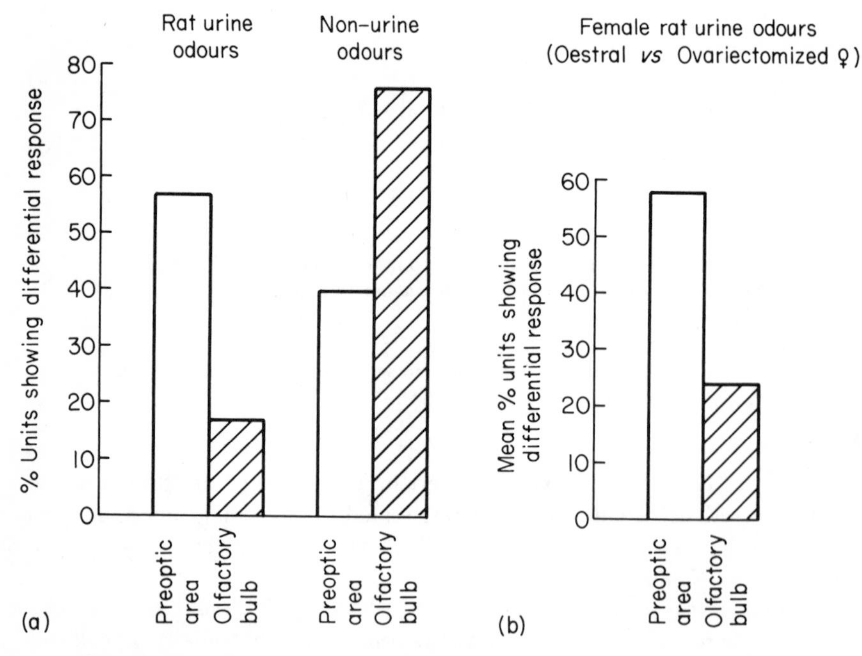

Figure 4.4 Histograms showing the percentage of units in the pre-optic region and the olfactory bulb of the male rat which respond differentially to
(a) rat urine odours and non-urine odours, and
(b) the odour of urine of oestral and ovariectomized female rats. (After Pfaffmann, 1971.)

Fur trappers, gamekeepers and others who regularly have to attract animals to traps frequently use odorous products of conspecifics. Seldom, however, has the undoubted effectiveness of treatment of the trap with the odorous product been subject to objective analysis. This is a pity, because the overall effect of captivity and a laboratory pedigree such as most laboratory animals bear has an incalculable effect on normal behavioural responses, and an experimenter can never be sure that a particular effect, clearly demonstrable in the laboratory, ever actually occurs in the field. The problem of field observation reaches somewhere near its maximum dimension when the subjects for study are the small, terrestrial mammals; the very species which are so intensively used in the laboratory. One field study, however, has indicated the potential importance of odours functioning as attractants. *Microtus townsendii* is a meadow vole from Western Canada which is easy to trap. If the number of individuals present on a study trapping grid is known from recapture studies, the effectiveness of clean and dirty traps can be compared. Fig. 4.5 shows that trappability during the breeding season is substantially lower when clean, as opposed to dirty, traps are used. In winter, after the breeding season, trap condition exerts little difference. Examining this relationship further, a correlation of 0.72 ($p < 0.005$) exists between female breeding condition and the magnitude of the drop in 'trappability' when a week in which clean traps were used is compared with a similar period during which dirty traps were used (Boonstra and Krebs, 1976). Clearly

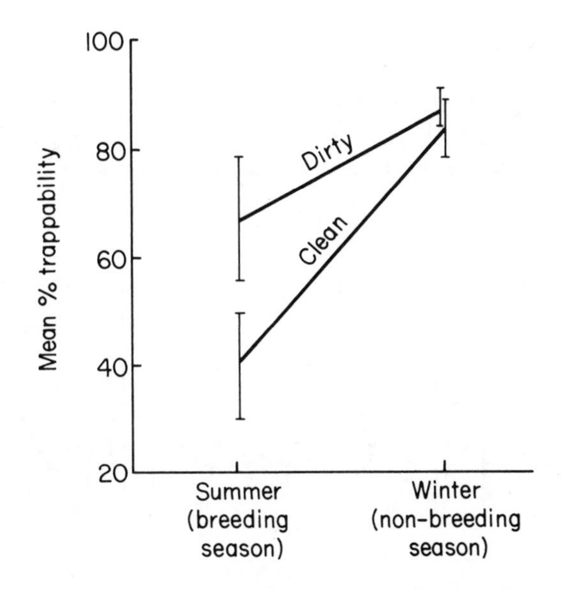

Figure 4.5 Mean trappability of *Microtus townsendii* in breeding and non-breeding season, using either all clean or all dirty traps. Vertical bars show 95% confidence limits. (From Boonstra and Krebs, 1976.)

the breeding males are attracted to the odour of the breeding females left behind in the traps. Dirty traps are literally just that. After having held an animal overnight they were roughly scraped out and reset. The specific attractive odour was not determined in this experiment; urine, faeces and lateral sebaceous organ secretions were all present in the dirty traps.

In stark contrast to these effects, which are at least partly caused by urinary odours, the urine of female golden hamsters is not attractive to males (Johnston, 1975). The vaginal secretions, however, are highly attractive and are copiously secreted during the oestrous phase of the cycle. The precise site of their production is not known; the upper vaginal wall is a specialized secretory epithelium, to the lower ventral aspect of the vagina lie the preputial glands and

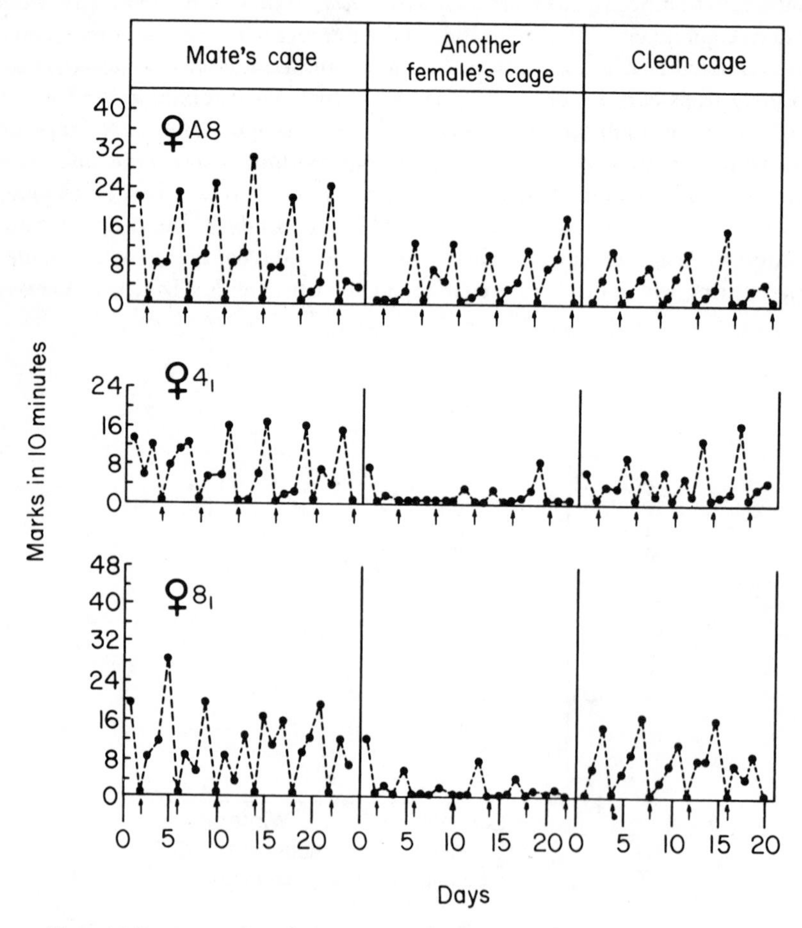

Figure 4.6 Frequency of vaginal marking by female golden hamsters in daily 10-minute tests in vacant stimulus cages. (↑) indicates female's oestral day. (From Johnston, 1977.)

the ducts to the urethral glands. Any of these sites may be involved, as may the bacterial decay of sloughed wall cells and other products (Johnston, 1977). Golden hamsters show a clear vaginal marking behaviour in which the orifice is extruded and dabbed onto the substratum. A cycle of marking frequency occurs with maximum frequencies the day preceding ovulation. Presence of a male, or even a male's cage, enhances marking frequency (Fig. 4.6). Male hamsters make quite appreciable efforts to get at the source of vaginal secretion, digging in the substratum and biting at an odour outlet pipe. They may spend nearly three minutes of a four-minute test period involved in sniffing the odour (Johnston, 1974). If a castrated male hamster is scented with vaginal secretions taken from an oestral female and put together with an intact male, the castrate is taken to be a receptive female and mounted (Fig. 4.7) (Johnston, 1977). Another observation highlighting the difference between mice and hamsters is that sexually inexperienced male hamsters, even those which have been isolated from females since weaning, are attracted to the odour of vaginal secretion. This is androgen-dependent, for castrates are not affected (Gregory, Engel and Pfaff, 1975). Although field studies are pitifully lacking, it is possible to explain these differences between hamsters and rats and mice in terms of population dispersion. Hamsters appear to live rather solitary lives; both sexes are highly aggressive and the well-known fact that they must be kept singly in laboratories suggests that they live singly in the wild. The well-developed vaginal marking behaviour and the potency of the vaginal secretion strongly suggests that female

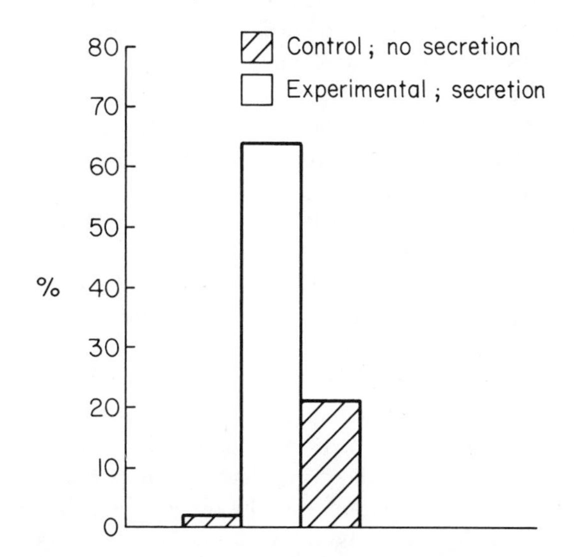

Figure 4.7 Percentage of trials in which an intact, adult male golden hamster, *Mesocricetus auratus*, attempts to mount a castrated male which has been either tainted with female hamster vaginal secretion or left untainted. (After Johnston, 1977.)

hamsters are subject to the same problems of mate finding as are, say, female cats. Broadcast of an odour signal attracts all potential mates from a considerable distance. There would be no selective advantage in attracting immatures, or others lacking threshold values of blood testosterone, which presumably would be incapable of carrying out successful mating anyway. Mice and rats, and presumably voles as well, are far more gregarious than hamsters and it would be advantageous for there to be some control over attraction and mating so that time and energy budgets are not needlessly disrupted. It is not known how sexually naïve males gain their first mating experience, or with which dominance level of individual they experience it. But once this behavioural and developmental hurdle is cleared, rather like an initiation rite, attraction to female vaginal secretion odour is strong.

In functioning as attractants for the opposite sex, odour signals are acting as advertisements of the sexual condition of the sender. We have seen that the most important message in the vaginal secretions is the state of oestrus of the producer, although it is not known why oestral secretions are more attractive than an- or dioestral secretions. It has long been known that the vaginal secretions of female rhesus monkeys (*Macaca mulatta*) become highly attractive to males shortly before ovulation, and that short-chain aliphatic acids are importantly involved in this attraction (Michael, Keverne and Bonsall, 1971). The proportions of acids change quite markedly during the cycle, with the total quantity reaching a maximum during the second quarter (Michael and Bonsall, 1977); the same occurs in women and, as also occurs in the rhesus monkey, acetic acid is the single most important of the straight-chain acids. The chief difference between the two cycles of acid production is that, in the monkey, butanoic is the next most abundant straight-chain acid and methylpentanoic the most abundant branch chain acid; in the human it is propanoic and methylbutanoic respectively (Michael, Bonsall and Kutner, 1975). There can be little doubt that the changing ratios in the monkey are responsible for at least a part of the advertisement and attraction; in the human no such effect has been experimentally verified (Doty *et al.*, 1975).

Advertisement of sexual status through odours may be very widespread throughout the vertebrates, and indeed it is very difficult to differentiate between sex attraction and advertisement. A curious feature of the biology of hedgehogs is that they anoint themselves with sticky saliva which, to the human observer, has a sharp, rank smell resembling horse's urine. The important point is that self-anointing occurs only in the breeding season and both sexes exhibit this behaviour. It is reported that the saliva contains an exudate from Jacobson's organ (Poduschka and Firbas, 1969), but this observation, and any behavioural effect it may mediate, requires to be checked. The odoriferous saliva may inform other hedgehogs of the presence and sexual status of another, although the occurrence of this behaviour in immature animals argues against a single simple interpretation (Brockie, 1976).

4.2 Detection and induction of oestrus, ovulation and lordosis

One of the major functions of courtship is to dissuade individuals from mating when there is a high chance that, because the female is not yet at the peak of her sexual receptiveness, the mating would be worthless. Thus the male requires to know precisely when his efforts are likely to be most successful. Many visual signals of readiness in fish and birds are known; there are few which unequivocally involve odours. There is one fascinating example of a fish perceiving the spawning readiness cues not of its own species but of another species. The green sunfish, *Lepomis cyanellus*, lays its eggs in the nest built by the redfin shiner, *Notropis umbratilis*. This latter looks after its eggs by fanning water over them – a procedure that likewise benefits the sunfish eggs. The odour of the milt and eggs produced by the male and female redfin shiner attract the male and female sunfish and induces them to spawn and fertilize their eggs. The developing sunfish eggs are abandoned by their parents to the gentle fanning and care of the shiner (Hunter and Hasler, 1965). In several species of fish a substance which inhibits spawning has been detected. The substance, which for goldfish at least is soluble in chloroform (Yu and Perlmutter, 1970), has a progressively inhibitory effect as the population density rises. If water in which a densely packed population has been held is presented to uncrowded ripe fish, their spawning is inhibited. A change of water removes the repressive substances (Swingle, 1953). This mechanism neatly prevents resource overexploitation.

The involvement of odour in this aspect of reproductive biology is particularly well-developed in the mammals. It has already been shown that odorous exudates change during the sexual cycle and that this change can be perceived (Figs. 4.4, 4.6 and 4.7). In the examples given, little is known of the mechanism whereby the state of sexual readiness is detected. The electrophysiological studies referred to on p. 94 indicate that the nose is the organ through which olfactory detection is effected. There is considerable evidence, however, to implicate the organ of Jacobson in this specific olfactory task, though there is by no means a consensus of opinion over this point.

There is neuroanatomical evidence supporting a hypothesis that Jacobson's organ is implicated in the recognition of sexual odours. The principal nerve from the organ is the vomeronasal nerve which passes directly to the accessory olfactory bulb. This structure closely resembles the main olfactory bulb to which it is juxtaposed at its posterior margin, but it is only a fraction of the size of the olfactory bulb. Neurons pass rearwards from the accessory olfactory bulb along the lateral olfactory tract to the mediocortical complex of the amygdaloid nucleus. Neurons from the main olfactory bulb pass to the olfactory tubercle but not, apparently, to the amygdaloid complex (Fig. 1.10) (Heimer, 1968). This suggests that the main olfactory and the vomeronasal systems provide separate, but parallel, routes of chemosensory influence into the hypothalamic region of the brain. Bilateral ablation of the medial pre-optic/anterior hypothalamic zone abolishes copulatory behaviour in the rat, proving that this region of the brain controls sexual behaviour. But the olfactory bulb does not project to the medial preoptic-anterior hypothalamic zone; it stops short at the posteromedial part of the cortical amygdaloid nucleus. Ablation of the olfactory bulb (and therefore also the accessory olfactory bulb) results in cessation of copulatory activity in the male rat, but the main olfactory bulb does not make contact with the hypothalamic region. It seems that the accessory olfactory bulb and its associated Jacobson's organ, and perhaps also the organ of Rudolfo-Masera, play an important though not equally fundamental role in regulating mammalian sexual behaviour (Winans and Scalia, 1970). Recent studies have shown that surgical ablation of the vomeronasal nerve of hamsters and guinea pigs results in very severe, but not total, reduction in the frequency of some aspects of reproductive behaviour (Powers and Winans, 1975; Planel, 1954). The whole story cannot be as simple as this, however, for not all mammals have a Jacobson's organ and accessory olfactory bulb. Higher primates lack Jacobson's organ and some bats, notably in the families Emballonuroidea, Rhinolophoidea and Vespertilionoidea, lack an accessory olfactory bulb. Whether sexual behaviour in these groups is mediated via the main olfactory bulb or some other, perhaps non-olfactory pathway, remains as yet an open question.

Other evidence comes from observations of 'Flehmen', a strictly mammalian behaviour characterized by an individual curling its upper lips as if it were expressing disgust at an unpleasant odour. The action of the lip-raising muscles, the *musculus levator labii*, exerts an upwards and backwards force on the anterior cartilaginous prongs of Jacobson's organ in such a way that the entrance to the incisive canal (i.e., the buccal end of the nasopalatine canal) is cleared. At the same time a slight decrease in pressure in the lumen of Jacobson's organ is wrought by the collapse of the vascular erectile tissue which runs alongside, and closely appressed to, the organ. This slight vacuum causes mucus-bearing dissolved odorous molecules to enter the canal and to pass into the sensory part of the organ. Expulsion of these particles and the replenishment with fresh is brought about by the pressure changes induced on the organ by the inflation and deflation of the erectile vascular bodies (Fig. 4.8). Flehmen has been reported to occur in a very wide range of mammals and is exhibited by both sexes whether they be fertile or not. It is neither androgen- nor oestrogen-dependent, as castrated male and speyed female cats show it. Flehmen in cats is only exhibited when one individual is in the presence of others of the opposite sex and almost always immediately after a stream or pool of urine has been investigated. Further experimentation to determine the relationship between reproduction and Jacobson's organ is difficult because of the difficulties involved in depriving the individual of the benefit of the organ. One attempt has been made to tackle this problem by (a) blocking the entrance to the organ with a silicone rubber plug; (b) injection of histoacryl into the nasopalatine canal; (c) suffusion of the organ by lavage with zinc sulphate solution; and (d) surgical mutilation of the

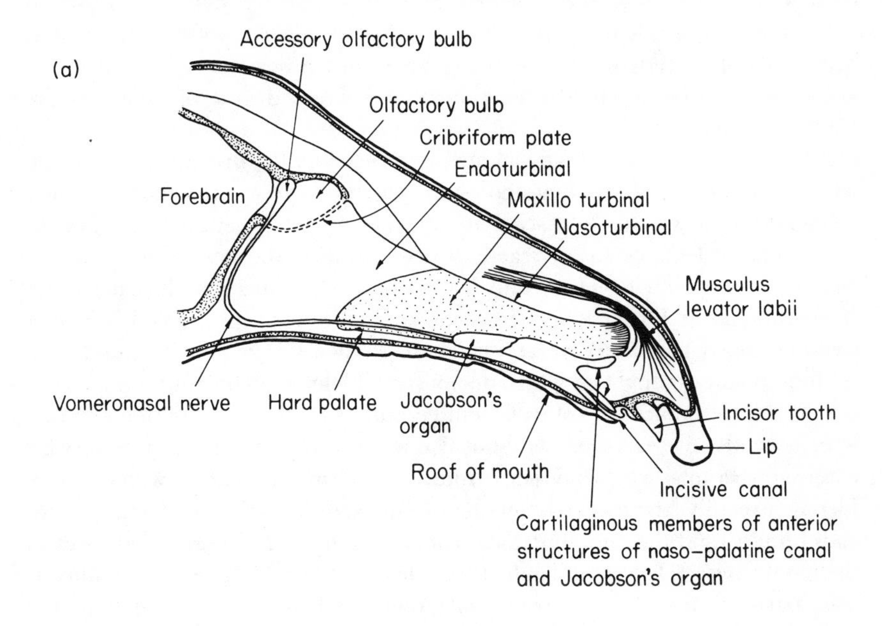

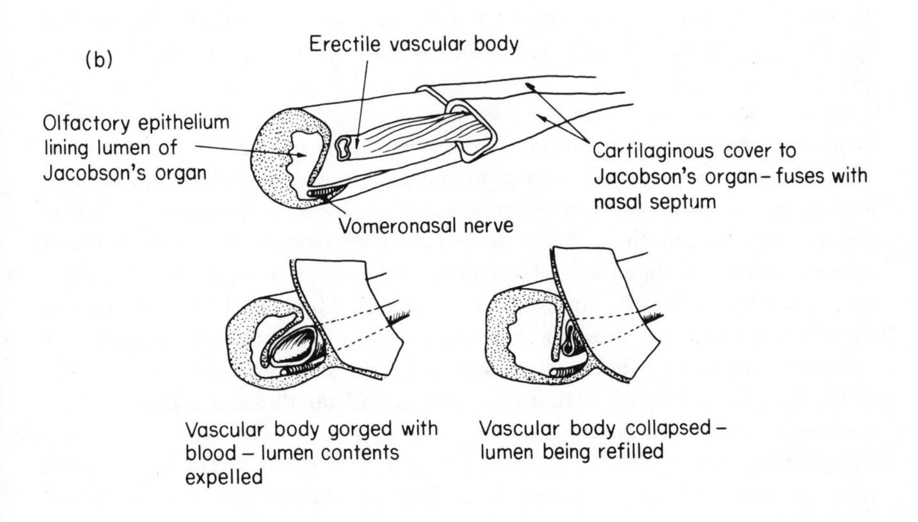

Figure 4.8 The mode of action of Jacobson's organ:
(a) (Based on Mann, 1961.)
(b) (From Mann, 1961.)

nasopalatine canal in an effort to cause sufficient deformation to block the passage. The result of these procedures is that the frequency of Flehmen behaviour was somewhat suppressed, but by no means abolished, and most experimental animals almost wholly regained their pretreatment rates within a few weeks (Verberne, 1976). This might indicate a degree of failure of the blocking technique.

It is significant to note that Flehmen is induced by the presentation of urine odours; non-urine odours have no effect. In those species in which Flehmen behaviour has been closely studied, it appears that its frequency is far higher in males than in females and it reaches its peak during the period in which the females are in oestrus (Dagg and Taub, 1970; Tomkins and Bryant, 1974; Verberne and de Boer, 1976). However, there are some species which have a well-developed Jacobson's organ, which are known to rely heavily on odorous stimuli in reproduction and which do not show Flehmen behaviour. One such is the pig which, on account of its lip musculature structure, no doubt associated with its ability to use its snout as a shovel, does not have the muscles necessary for effecting the lip-curl behaviour. Additionally, in the horse, which shows Flehmen readily, the nasopalatine duct from Jacobson's organ into the mouth ends blindly, and the only functional connection from the organ is a duct from the nasal cavity (Dagg and Taub, 1970; Negus, 1958). Flehmen is probably at least partly influenced by normal olfaction, for it appears that it is totally abolished by ablation of the olfactory bulbs in rams (Lindsay, 1965). So, while there is some circumstantial evidence of, and inferred implication for, the role of Jacobson's organ in the detection of the state of sexual readiness of a partner, the specific functions of the organ remain somewhat obscure (Knappe, 1964; Shank, 1972).

Mammals cannot mate until the female has adopted a specific mating stance or posture; the process and result of adopting the correct stance being termed 'lordosis'. For the huge majority of mammals, we do not know what induces lordosis, although in antelopes sight and touch are known to be important. In the pig, odours derived from the 16-androster androgens induce the 'standing reaction' in the sow, in which a rigid stance is adopted with the ears held straight up. Only sows in oestrus exhibit the reaction and, when no boar is present, less than half of them will show the reaction to slight back pressure applied by a human observer. If a boar is present which can neither be seen nor touched, 90% of the sows show lordosis. Addition of visual and tactile stimuli increases the number of females responding by 7% and 3% respectively. One of the most important active ingredients in boar odour is 5α-androst-16-ene-3-one, and this is as effective in eliciting the reaction as is the fluid drawn from the preputial pouch. It is important to note that olfactory signals alone are not quite as effective as the close proximity of an intact boar (Signoret, 1976). A contradictory situation occurs in rats, however. Sexual receptivity, that is the proportion of female rats which adopt lordosis and accept the male, is enhanced by the removal of the olfactory bulbs of the female (Satli and Aron, 1976). Nothing is known about the basis of choice of male by the females of these species, so it is possible that female rats are able to discriminate between less desirable and more desirable males by their noses. When deprived of this ability, they accept all comers. What happens under natural conditions, where dominance hierarchies are strong, is not known.

In several species of mammals the odour of the male is able to induce oestrus in the female. Presumably this is of selective advantage to males of gregarious species, so that a small number of males can mate with a large number of females, and it would clearly not be of advantage to solitary species in which the physical dispersion of males and females is sometimes very great. In these species, space considerations alone dictate that any one male cannot mate with more than one female. In the mouse, the odorous substance contained in the urine of the male is small enough to be carried at least eight feet in an air current of just 0.4 km/h (Whitten, Bronson and Greenstein, 1968). Although the identity of the oestrus-inducing substance is not known, urine removed from the bladder of mice is just as effective as naturally voided urine, suggesting that exudates from the preputial glands are not involved. Castration of urine-donating males results in loss of the oestrus-inducing activity of urine, suggesting that the active ingredient is either an androgen metabolite or the product of some tissue maintained by androgens. The urine of female mice implanted with testosterone also bears this substance (Whitten and Bronson, 1970; Bronson, 1971), which seems to implicate the hormone beyond doubt. Only very small amounts of male urine are required to increase significantly the proportion of females entering oestrus. As little as 0.01 ml of urine dripped into a cage containing 8–10 adult female wild house mice or deer mice, *Peromyscus maniculatus*, is sufficient. The response is species-specific; male deermouse urine is ineffective with house mice and vice versa. Fig. 4.9 summarizes this situation as revealed by analysis of vaginal cell types (Bronson, 1971). If, however, the male itself and not just its urine is introduced into the cage of females, subsequent observation of the timing of births gives a clue as to what is actually meant by 'oestrus induction'. If female mice exhibiting normal oestrous cycles are removed from stock boxes and housed on their own in an environment free from the odours of males, their cycle lengths increase and the frequency of pseudo-pregnancy rises. Eventually the majority of females enter anoestrus (Van der Lee and Boot, 1955). If a male is now introduced into the experimental cage, a sharp peak in parturition is observed after a duration of three days longer than the normal gestation. If just the urine of an adult male is introduced for two days, followed by the male itself on the third, the peak of parturition follows after a delay exactly equalling the length of gestation. Studies on the vaginal wall cell types reveal that the odour of the male brings the rather vague and lengthy oestrous cycles into *synchrony* with one another. Sexual receptivity reaches its peak on about the third day of a normal four- to five-day cycle – hence the peak of births occurring at three days longer than the normal gestation. Fig. 4.10

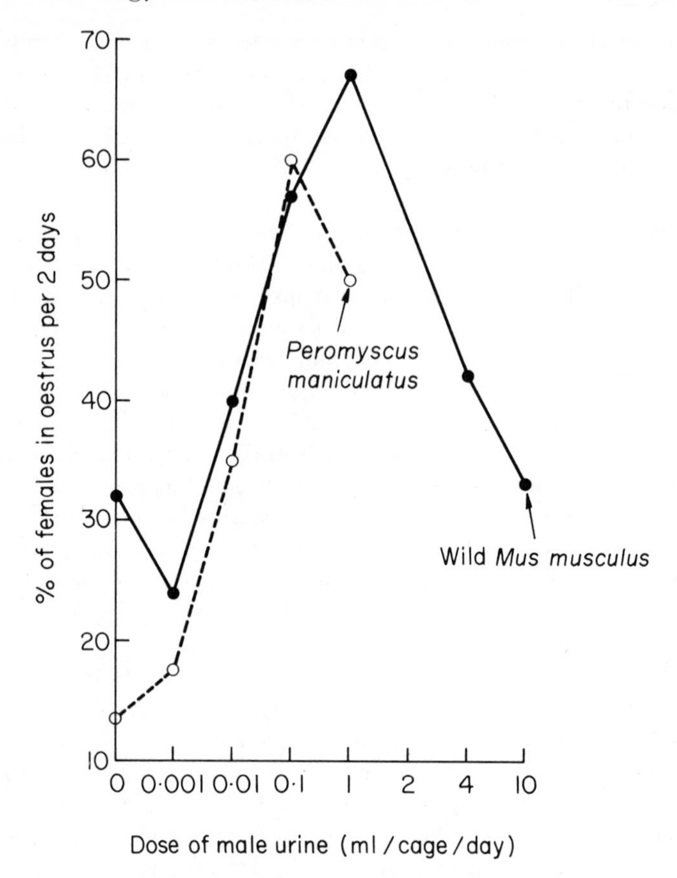

Figure 4.9 The induction of oestrus in *Mus musculus* and *Peromyscus maniculatus* by the application of male urine to females' cages. (After Bronson, 1971.)

summarizes the results of many experiments designed to examine this phenomenon (Whitten, 1956). It is clear, then, that induction of oestrus is really the speeding up and bringing into line of cycles which apparently require the presence of a male-produced timing cue. How the odorous substances in the male's urine actually influence the oestrous cycle is not known, but presumably the impulses generated in the nose and transmitted to the pre-optic area influence the release of gonadotropin by the anterior pituitary via the hypothalamus. Although such a dramatic response to male odour as that shown in Fig. 4.10 would be inconceivable in a naturally occurring rodent population, the action of the males would be to maintain the reproductive status of the females in the population. The same substance which induces lordosis in the sow is responsible for influencing the return to oestrus after parturition. Table 4.3 shows that treatment of the sow two or four days after weaning causes a significantly earlier

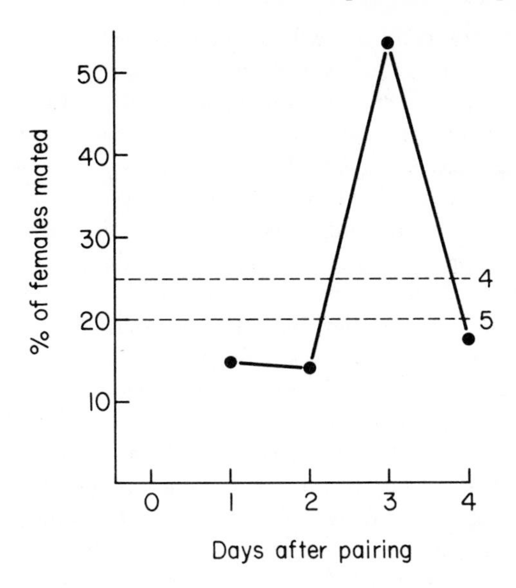

Figure 4.10 Graph to show percentage of female mice mated each day for four days after the introduction of stud males. Broken lines show proportion of matings occurring each day in mixed sex groups with four- and five-day oestrous cycles. (After Whitten, 1956.)

return to oestrus than occurs with untreated sows (Hillyer, 1976). Presumably under natural conditions the omnipresence of boars would render the 'control' situation in Table 4.3 irrelevant.

Table 4.3 The effect of synthetic boar odour substances on resumption of oestrus after weaning in sows (from Hillyer, 1976).

	One- or two-second spray of synthetic steroid* on the nose		
	Two days after weaning	*Four days after weaning*	*Control*
Number of sows	34	71	92
Mean time (days) from weaning to conception	10.3	9.0	27.2

* 'Boar Mate' made by Jeyes Animal Health Division, U.K.

There are enough observations about the effect of the presence of a male on the onset and synchrony of oestrus in females of domestic stock to suggest that it is likely that natural selection has permitted the development of a system for socially induced oestrus control in wild ungulates. Sheep breeders know that the presence of a ram in a group of ewes will not only bring the oestrous cycles into synchrony, but will also hasten the onset of oestrus by up to 50% (Watson and Redford, 1960; Parsons and Hunter, 1967). Similar observations come from studies of domestic goats, *Capra hircus*, and from nutria, *Myocastor coypus* (Shelton, 1960; Ehrlich, 1966). Males of very many, and probably most, species of ungulates indulge in the spreading of their own urine all over their bodies in the period immediately prior to mating, during the breeding season and for a short while after it. Goats urinate directly onto the underside of the body and even over the face. Bovids and cervids urinate onto the ground, then wallow in the pool so formed and cover the flanks and back with a mixture of urine and soil. Camels, *Camelus dromedarius*, urinate onto the end of the tail which is then slapped against the flanks alternately. Reindeer, *Rangifer tarandus*, and New World deer, Odocoilinae, urinate onto the tarsal region of the hind legs (Coblentz, 1976). Self-anointment with urine is thus a behavioural commonplace in wild ungulates. While there is no overall agreement about the function of this urination, indeed it is not known whether it is primarily for inter- or intrasexual use, it is at least possible that it functions to bring the females into oestrus all together, thereby enhancing the chances of the fittest males to mate with as many females as possible.

4.3 Courtship, mating and related behaviours

Courtship functions chiefly as a safety net to preclude either different species mating or a behaviourly defective individual from mating. It is frequently very complex in the mammals, involving many hours, or even days, of pursuit, mock attacks and appeasements. At any stage if the wrong cues or responses are given by either partner the intended mating does not occur. There is little difference between the procedure of courtship in species having internal or external fertilization. A whole range of courtship and mating signals are known in fish, from the purely visual attraction by the female stickleback, *Gasterosteus aculeatus*, for the male, through the cutaneous anal gland secretion of the male *Blennius pavo* which attracts the female, to the secretions of the gravid ovary of the goldfish, *Carassius auratus*, and goby, *Bathygobius soporator*, which attract the male to the female. Plugging the nostrils of male *Bathygobius* results in elimination of courtship behaviour, indicating the importance of the odour cue. Courting males themselves exude chemical signallers which arouse other males into courtship in at least one genus, *Hypsoblennius*, but the influence of sight and acoustic signals on olfactory signals has yet to be determined (Tavolga, 1976). In the goldfish, it is known that the blood androgen level of the male is a key factor in the elicitation of prespawning courtship; sexually immature males show little interest in gravid females (Partridge, Liley and Stacey, 1976).

There is clear evidence, then, that courtship in fish leading to fertilization depends to some extent upon the production of odours either by the male, female or both. As was noted in the preamble to this chapter, mating in mammals is seldom totally eliminated by loss of olfactory power, although it is often seriously impaired. For example, intact rams usually inseminate almost all ewes (97%) that are in oestrus at any one time, but rams deprived of their olfactory bulbs by surgery only inseminate a little over half the ewes in oestrus (55%) (Fletcher and Lindsay, 1968). Suffusion of the olfactory membranes with zinc sulphate solution causes temporary anosmia and rams treated in this manner mate with oestral ewes 4.38 times less frequently during a 10-minute period than control rams (Rouger, 1973). Neither of these treatments succeeds in totally eliminating mating behaviour in sheep, but in some species total elimination does occur. One such is the golden hamster, *Mesocricetus auratus*. Zinc sulphate suffusion totally suppresses all courtship and copulatory behaviour and males so treated show no sexual interest whatsoever in normal females (Lisk, Zeiss and Ciaccio, 1972). Such total dependence may have a selective advantage in this species in which the female dominates the male, and may inflict serious harm on him should a fight break out during a clumsy courtship. Most rodents are able to mate in spite of olfactory impairment, though invariably reproductive performance is significantly weakened. In the laboratory rat, for example, all stages of mating show some impairment. Anosmic rats often mount normal females and achieve intromission, but frequently do not ejaculate. For those that do, the latency to the next ejaculation is significantly longer than in controls (Larsson, 1971).

As far as reptiles in general are concerned, it is not known whether olfaction plays any more than a minor part in courtship. In some species, however, it fulfils a clearly defined role. During the early stages of courtship in the tortoise, *Gopherus polyphemus*, the female rubs her forearms against her chin gland. She then repeatedly proffers her forearms to passing males who may take up the offer and investigate her more closely (Auffenberg, 1966). During the breeding season, females of most species of snakes produce a strong odour from a secretion expelled by the cloacal gland. In *Vipera berus* some females are more attractive than others; possibly they exhibit a sort of oestrus which renders them maximally attractive. The odour of the cloacal secretions are quite species-specific. If a male garter snake, *Thamnophis butleri*, is courting a female *T. butleri* and that female is exchanged for a *T. sirtalis*, the male *T. butleri* approaches the new snake, flicks its tongue a few times and then turns away. Visually there is little to tell the two species apart, and although subtle visual cues may be involved in the courtship, olfactory cues seem to dominate. In other species of snake, *viz*: the timber rattler, *Crotalus horridus*; the copperhead, *Agkistrodon mokasen* and the water moccasin, *A. piscivorus*, the odour of the whole body surface, and not just the cloaca, serves to attract males to females and induce courtship (Noble, 1937).

In large wild mammals, the role of olfaction in courtship and mating can only be assessed by careful field observation. Seasonal production of odorous material from the temporal glands of elephants has long been recognized to be related to breeding. An active temporal gland is indicative of the state known as 'musth', and is only seen in males. It is never seen in young males less than 10 years old, sporadically in males between 14 and 20 years, and regularly in males over 20 years. It occurs in relation to the rut, which coincides with the wet season. A bull in musth rubs its trunk on the temporal region of its face picking up much of the sticky secretion. It then smears this on tree trunks, around its mud baths and upon its own back and sides. The production of musth is not physiologically necessary for reproduction to occur, as breeding sometimes occurs outside the period of temporal gland activity, but it appears to act as both an attractant to cows as well as an initiator of courtship. Cows are very interested in musth and sniff it whenever possible (Fig. 4.11), both from the male itself and from the places upon which it has been daubed (Eisenberg, McKay and Jainudeen, 1971). Similar scent production from cephalic glands associated with the rut is known in the camel; once again successful mating can occur outside the period of odour production, but presumably under totally natural conditions such matings are less successful than when mate selection and courtship occurs in the presence of the odours (Pilters, 1956).

Scent urination in male ungulates has already been mentioned as a mechanism for quickening the onset of oestrus. It has been suggested further that urine acts primarily as a carrier of information about the physical state of its producer. If a male quickly becomes out of condition during the breeding season, and in many species the bulls do not feed during quite long periods, the changed pattern of metabolites in the urine will reflect the changed physiology of respiration. It would be of selective advantage for a rutting male to advertise his rugged constitution for as long as possible, so that minimal time is diverted from his courtship and copulatory activities. If, further, the odour of his urine has quickened the onset of, and synchronized the oestrous cycles of the females, courtship and mating of very many females must occur within a short time span. Any adaptation allowing him maximum time for these activities must have high selective advantage. But if his flagging physical condition is advertised to other, less dominant males, his deposition cannot be long awaited. The urine of the largest and strongest billy goats has, to the human nose at least, a more pungent and stronger odour than the urine of smaller, weaker males. If this stronger odour is able successfully to mask the odour associated with deteriorating condition, males producing it will be able to continue their mating activities for longer than if this deception had not occurred. Fig. 4.12 indicates the advantage to be gained by this deception (Coblentz, 1976). Selection for deception in matters of social behaviour is quite common among vertebrates; for example, there are many instances known in which tufts of hair or the inside convolution pattern of the ear have come to resemble extra horns, or extensions to other visual ornaments (Guthrie and Petocz, 1970). Olfactory deception is just as likely.

In very many species, mating is associated with a scent-marking behaviour in which the male places a scent mark on the female. This has the obvious selective

Figure 4.11
(a) Wild bull Asian elephant, *Elephas maximus*, in 'musth'. The free flow of secretion from the temporal gland is readily seen (arrowed).
(b) A female Asian elephant (on the left) sniffing the temporal gland of a bull in 'musth'. (Photographs: G. McKay.)

advantage of deterring other males from wasting time and energy on females already mated and is seen in a wide variety of species of mammals. In the

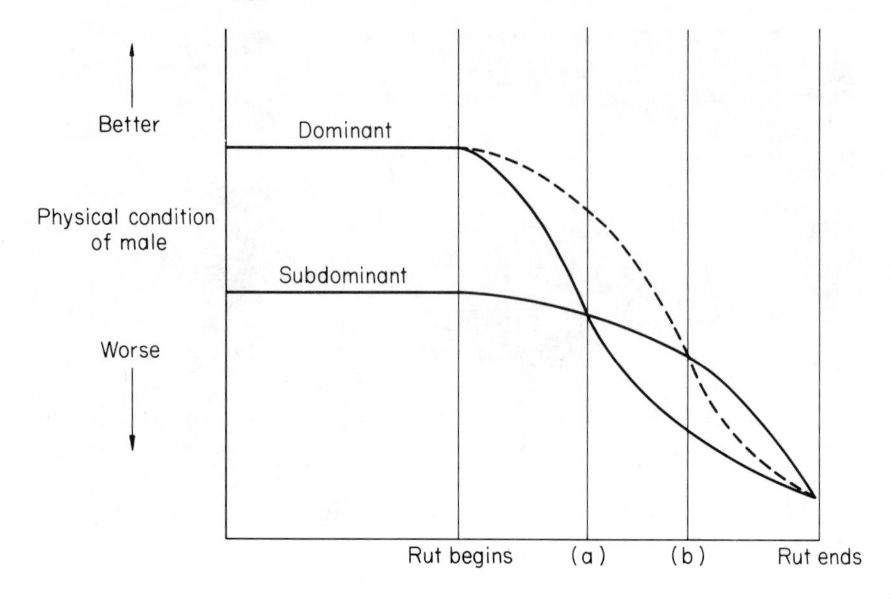

Figure 4.12 Hypothetical situation depicting the actual decline in physical condition of a dominant male caprid and the appearance to subdominants of his decline as evidenced by masking odours in the urine. Although the physical condition of the dominant fell below that of rival subdominants at (a), he was deposed only at (b).

marsupial sugar glider, *Petaurus breviceps*, and possibly other species with a sternal gland, the male applies some secretion to the head and the back of any female mated (Schultze-Westrum, 1965); the male rabbit, *Oryctolagus cuniculus*, porcupine, *Erethizon dorsatum*, and some other species urinate on the female prior to mating, and the rabbit, the steenbok, *Raphicercus campestris*, and other ungulates with an intermandibular gland anoint the female with the products of this gland during mating (Cohen and Gerneke, 1976). These actively applied marks are additional to the odours emanating from the seminal fluid and other passively applied marks resulting from mating.

4.4 Pregnancy

One of the most dramatic examples of the force of olfaction in reproduction stems from the late Hilda Bruce's observations that if the stud male is removed from the cage of an inseminated female mouse and his place taken by another adult male, or just his urine, the female's blastocysts fail to implant, they are resorbed and she quickly returns to normal oestrus (Bruce, 1950; Bruce and Parrott, 1960). If the females are rendered anosmic, no block to pregnancy occurs; it is the ability of the female to discriminate between the odours of two males which is responsible for the events to occur. Odour-induced pregnancy block has been recorded for *Peromyscus maniculatus*, *Microtus agrestis*, *M. ochrogaster* and *M. pennsylvanicus*, as well as the house mouse (Bronson, 1971; Clulow and Clarke, 1964; Stehn and Richmond, 1975). For long it was thought that pregnancy could only be blocked prior to implantation, for after five days female mice are no longer affected by the presence of strange males; but studies on *M. ochrogaster* have shown that the period of susceptibility to pregnancy block extends beyond the time of implantation and for up to 15 days of gestation. The introduction of a strange male during this period causes pregnancy termination in over 80% of females. Between 15 and 17 days (normal gestation is about 25 days) the percentage of females aborting drops to 36%. Those pregnancies maintained in spite of the presence of the strange male are significantly shorter than control pregnancies, the litter size is smaller, and a smaller percentage survives until weaning. Although it is hard to see how it could be demonstrated, male-induced pregnancy block has yet to be shown to occur in wild populations. Presumably, none of the members of one population would cause termination, but immigrants from other populations might. Thus the establishment and maintenance of stable social conditions within rodent populations may be necessary for rapid population increase. It is perhaps significant that emigration and immigration in rodent populations tends to occur, though not exclusively, at the end of the breeding season. Since a strange male rodent, of high abortigenic calibre, can mate with the female whose litter he has caused to abort and can successfully bring a new litter to term, it would seem there is no *absolute* quality of urine odour which effects the endocrine change. Whether his urine odour rapidly changes to resemble that of his adopted population because he is feeding on their food stuffs, or whether his odour is quickly learned by his new-found conspecifics, is not known. Whatever happens, his ability to disrupt pregnancy rapidly vanishes. An odour-induced blockage to pregnancy has not been observed in mammals other than rodents.

Reference had already been made to the odours which induce lordosis in the sow. Because of the great interest which has recently been shown in the odour environment of breeding sows, attention has been paid to the quality of pregnancy resulting from artificial insemination. Significantly more sows which are subjected to odorous steroids bear litters (65.9%) than those sows not subjected to the steroids (46.4%). The litter size of 10.2 is somewhat higher than the 9.7 for untreated sows (Signoret and Bariteau, 1975). While these observations might be open to interpretation in a different way, they indicate that pregnancy is more common and more successful in those sows subject to a normal odorous environment than is usual in situations of artificial insemination. Rebred sows, that is, sows allowed to breed soon after the end of the normal 35-day suckling period, which are treated with a spray of artificial steroid at two or four days after weaning, produce piglets slightly larger (13.4 kg *vs* 12.7 kg) than control sows not treated with steroid (Hillyer, 1976). While this difference is not in itself statistically significant, on a large enough pig-breeding unit it may be economically significant.

4.5 Parental behaviour; imprinting

Parental behaviour is any behaviour pattern exhibited by one or other of the parents of a newborn or newly hatched youngster and can occur before birth or hatching. For the placental mammals, and particularly those with precocial young, a parental task to be completed prior to birth is the construction of a suitable nest. Not only must this be built in a place safe from predators but it must have the correct thermal qualities. Under laboratory conditions female mice will build a parturition or brood nest of cotton wool and, for an investigation of parental behaviour, it is a simple matter to arrange a hopper above the cage to supply cotton wool which can be removed daily to have its contents weighed. Pregnant female mice which have undergone surgical olfactory bulbectomy pull very much less cotton out of the hopper for nest building than do either sham-operated or intact control females (Fig. 4.13). Progesterone is thought to be the hormone controlling nest building in the mouse, since therapy with this hormone sharply increases nest building in virgin females. Olfactory bulbectomy results in atrophy of the ovaries, and hence reduction in progesterone output. Therapy with progesterone should, if this explanation of the mechanism is correct, cause an increase in nest building in mice lacking their olfactory bulbs. Experiments have shown that this does occur, but such treated mice never pull as much cotton wool as sham-operated controls. The full role of the olfactory bulbs in this behaviour awaits a further analysis (Zarrow, Gandelman and Denenberg, 1971; Whitten, 1956).

For parental care to be effective, the parents of newly hatched or newborn young have to be able to discriminate between their own and the young of other parents. Such a problem only applies to colonial breeders, naturally enough, although there is no evidence that strongly territorial species are any less good at recognizing their own young. The process whereby the parent becomes aware of the particular characteristics of its own young, and *vice versa*, is called 'imprinting', and it occurs at the earliest possible moment in the free life of the young. The odour of mammalian birth fluids themselves, some of which are voided prior to birth, allows parturient females to know their own young immediately they are born. Under normal conditions soay sheep ewes show a lip-licking behaviour shortly before the birth of their young, followed by intense licking of the lamb as soon as it is born. Within half an hour the lamb is free from

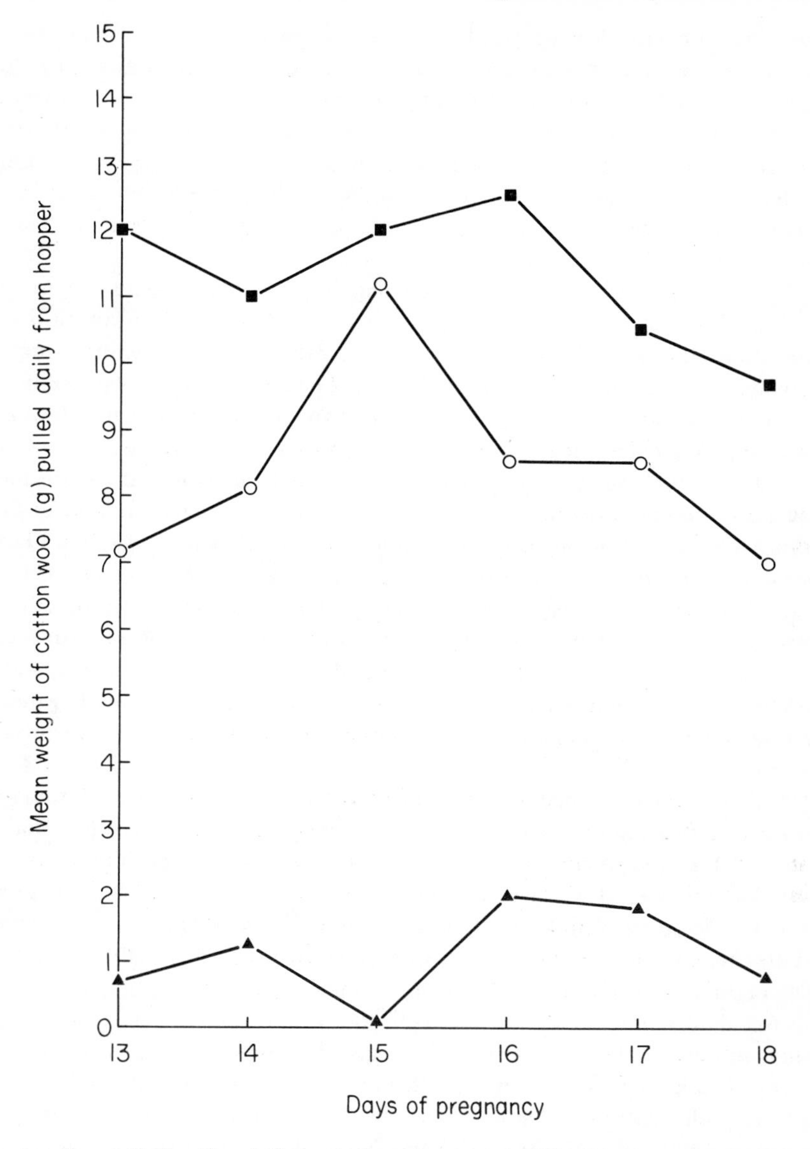

Figure 4.13 The effect of olfactory bulbectomy on nest-building behaviour of pregnant female mice. X–X control group (n = 10); o–o sham-operated group (n = 10); ▲–▲ bulbectomized group (n = 10). (From Zarrow, Gandelman and Denenberg, 1971.)

caul and birth fluids and is almost dry. Pregnant ewes rendered anosmic by surgical removal of the olfactory bulbs show no lip-licking behaviour prior to the birth of their young and no licking of the newborn lamb afterwards. Such lambs remain wet and cold for a long time until the birth membranes gradually rub off.

The difficulty of cross-fostering orphan lambs on ewes with stillborn young is well known; shepherds have to dress the orphan in the skin of the dead lamb so the foster mother may be fooled into accepting the orphan. With anosmic soay ewes cross-fostering is commonplace, and while young lambs make some effort to feed only from their own mothers, their anosmic mothers readily allow any lamb to suckle. The bond which links mother to young, with a force which sometimes defeats even the most experienced shepherd, has a firmly rooted olfactory basis (Baldwin and Shillito, 1974).

A moment's reflection on this rooting will underscore the fundamental basis of olfaction. Most mammals–ungulates are exceptional–are born in an altricial state, often with their eyes and ears largely inoperative. But the olfactory system, partly on account of its functional simplicity and partly on account of its position within the respiratory system, is well-developed and functional at birth. The role of olfaction in guiding the newborn marsupial, which has no eyes nor ears, from the cloaca to the pouch, has already been described. As soon as it is born the young mammal must breathe and, because of the juxtaposition of the olfactorily receptive tissue and the air passage, it must stimulate its olfactory system. For very many, and perhaps all, species the first activity of the young is to seek a nipple and attempt to feed. The mammary glands themselves are modified sebaceous glands and their nipples are frequently surrounded with banks of holocrine sebaceous glands adding nothing to the milk. The first social contacts, therefore, of a newborn youngster are associated with odorous tissue. It would seem reasonable to expect that olfactory imprinting in mammals is of universal occurrence.

Olfactory imprinting occurs very quickly. If a young goat is removed from its mother immediately after it is born and before the mother has made any contact with it, and is subsequently returned to the dam after having been thoroughly washed and cleaned of all traces of embryonic membranes and fluids, the mother actively rejects it by butting. But if the kid is left with its mother for as little as five minutes before being removed and washed, she will accept her kid even after a delay of three hours (Klopfer, Adams and Klopfer, 1962). Imprinting serves not only to aid the mother in identification of her own young but also aids the young to find its own mother. It is as important that the young of gregarious species should be able to find their own mothers as it is for the mothers actively to discourage the suckling advances made by alien young. For this reason, the critical period during which olfactory imprinting occurs is much briefer than in solitary species which breed in secluded and isolated places. In many social rodents the period is extremely brief; as short as one hour in the precocial murid, *Acomys cahirinus* (Porter and Etscorn, 1974). While the existence of the critical period serves to ensure that each youngster is given the best chance by its own mother, it is not quite as immutable as it seems. If newborn lambs are put with newly parturient nanny goats who have been allowed to make contact with their own newborn kids, or newborn kids with ewes, the females attempt to butt the foreign youngster away. But if the adult is forcibly restrained, she gradually comes to accept the young stranger, though it may take more than 10 days for the butting behaviour to cease. Thereafter there are few problems; she will suckle the adoptive youngster as readily as she would her own (Hersher, Richmond and Moore, 1963). This indicates that the species specific template is not genetically determined; rather the occurrence of a short period when the odour memory is highly receptive and finely tuned allows the mother to respond in a favourable way to only her own young. The fact that the critical period can be extended under artificial conditions is further evidence of the lack of genetic control of olfactory imprinting.

As with young birds imprinted for life upon cardboard boxes or human beings, the early olfactory experience of mammals exerts a strong influence on social preferences later in life. Not all species show the same degree of dependence upon olfactory cues, however. Guinea pigs, *Cavia porcellus*, can be visually imprinted upon wooden blocks, but attempts to imprint them upon ethylbenzoate or acetophenone fail after a few weeks. (Carter and Marr, 1970). Young rats, on the other hand, reared with mother and litter mates characterized by the odour of ethyl benzoate are later, as adults, much more responsive to sex partners smelling of ethyl benzoate than untreated control partners (Marr and Gardner, 1965). There can be little doubt that this difference is related to the peculiar socio-ecological conditions in which these two very different species live.

A curious feature of rodent-breeding biology, which strongly contrasts with that of ungulates, is that, although odour cues play an important role in social interactions, females which have recently given birth will readily accept foster pups not only of their own species but also of other species. Foreign young can be introduced to a lactating female without problem during the first half of lactation and possibly beyond. If young grasshopper mice, *Onychomys torridus*, are fostered on deer-mice, *Peromyscus leucopus*, and vice versa, the young after weaning choose to spend more time in the vicinity of soiled bedding of the foster parent species than the natural parent species. Such a choice is maintained throughout life. Unfortunately, the studies which have been performed on cross-fostering have been abandoned before the mating preferences of cross-fostered young could be ascertained, so it is not certain how complete the imprinting was (McCarty and Southwick, 1977).

The young of most, but not all, ungulates and other precocial young mammals keep close to their mothers during her bouts of feeding and other activities. For seal pups, whose mothers return to sea each day to feed, close contact at all times is not possible, yet foster feeding in seals is a rare occurrence, although it is becoming increasingly common in the grey seal, *Halichoerus grypus*, for reasons associated with changing ecological conditions. Cow Alaskan fur seals, *Callorhinus ursinus*, and Weddell seals, *Leptonychotes weddelli*, make no attempt to lick the birth detritus from their newborn pups, but they do sniff their youngsters very thoroughly. When they return to the breeding colony from feeding at sea, they appear to find the right part of the colony where their own young lie by vision but make the final identification by nose-to-nose contact. All pups actively sniff the cows' muzzles, but all except the cows' own young are chased away (Mansfield, 1958; Bartholemew, 1959). Experimental proof that identification is based on olfaction came accidentally from growth and development studies on Weddell seals. Pups were weighed by being put in a sack; but the trauma of this usually caused them to defecate. If a pup was put into a sack containing faeces of another, it was subsequently rejected by its mother. If a clean sack were used for each weighing operation, no rejections occurred (Kaufman, Siniff and Reichle, 1975).

Parental care is well-developed in several families of fish, particularly the mouth breeders (Cichlidae). In this family the female keeps the brood of hatchlings in her mouth, except when they are feeding. Cichlids are carnivorous and adults usually feed upon prey items which closely resemble their own young in both size and movement pattern. Although the precision of brooding in cichlids is influenced by the female's past brooding history, there is clear evidence that the brood of hatchlings is distinguished from food by a chemical substance associated with the fry. If water in which a brood of young jewel fish, *Hemichromis biniaculatus*, have been swimming is introduced into an aquarium in which the parents are kept, the adults orientate towards the point of inflow and show fanning and other parental behaviour. Parental behaviour normally continues for two to three weeks but it can be lengthened by replacing older offspring with younger ones and curtailed by replacing young fry with older. Parental behaviour in the female is not released, even in an old female having had several broods previously, if the brood is suspended in her tank in a sealed but transparent glass jar. Olfaction remains the dominant sense when the fry are very young and while they are 'wiggling' rather than swimming; but later, when the youngsters are swimming properly, vision appears to become dominant. The presence of certain chemical cues is essential for the smooth switching of primary sensory modalities (Kühme, 1963, 1964; Myrberg, 1966).

The release of maternal care in mammals is dependent upon a functioning olfactory system. Mother mice from which the olfactory bulbs have been removed show no maternal care and eat their own young. While therapy with progesterone increases the amount of nest-building behaviour in bulbectomized females, it has no effect on maternal behaviour (Table 4.4). To what extent the induction of normal maternal behaviour depends upon the prior execution of nest building or other prebirth behaviours, and not primarily upon blood hormone titres, is not known.

Young rodents grow very quickly; within about 30 days of being born, at a very undeveloped stage, they are active, fully weaned and capable of looking after themselves. During the latter part of suckling it is encumbent upon the mother to exert a restraining influence upon her young to keep them safely marshalled within the nest and, hence, out of danger. Young rats and mice

Table 4.4 The response of female mice to their litters following surgical olfactory bulbectomy, sham operation, and progesterone and cholesterol benzoate implant therapy. (Modified from Zarrow, Gandelman and Denenberg, 1971.)

Group		% that: *Behaved maternally*	*Ate young*
Bulbectomized mothers	10	0	100
Sham-operated mothers	10	90	10
Control mothers	10	90	10

Group	*Hormone or placebo therapy*	n	% that: *Behaved maternally*	*Ignored young*	*Ate young*
Bulbectomized	Progesterone	10	0	20	80
	Cholesterol	10	0	10	90
Sham-operated	Progesterone	10	100	0	0
	Cholesterol	10	90	10	0
Control	Progesterone	10	90	0	10
	Cholesterol	10	90	10	0

removed from their nest are quickly found and retrieved by their mother who is guided in this task by both ultrasonic and olfactory cues. Her response is far quicker to her own young than to alien young (Fig. 4.14). Under natural conditions, however, it is unlikely that this retrieval mechanism would often be invoked. Instead, the mother produces an odour which is attractive to the young. The odour is produced in the caecum and taints not only the body of the mother, and thus everything she touches, but also her faeces. The odour originates in the bacterial flora of the caecum and treatment of the mother's gut with antibiotics suppresses production of the substance. Prolactin plays a role here, for bile fluid from a lactating rat (of lactational age of 21 days) induces the caecum of male rats to produce the attractive substance. Bile from newly parturient female rats, whose blood carries little prolactin, does not induce production of the substance in male rats. Maternal pheromone is not family-specific; any young rat will be attracted to the caecal odour of any lactating female. But it is influenced by the age of the litter. Normally disappearing at about 27 days, the substance will continue to be made for far longer if the female is given foster litters of a younger age; in this situation it is clear that the main governing factor in attractive

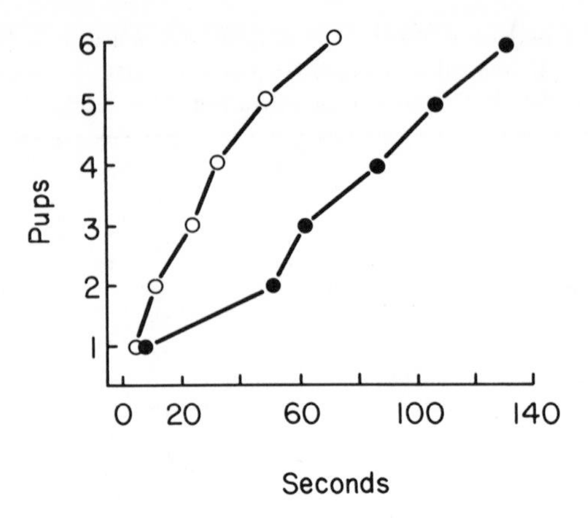

Figure 4.14 The length of time taken for a female mouse to retrieve six of its own *vs* six alien pups. o – o own pups; ● – ● alien pups. (After Beach and Jaynes, 1956.)

substance is lactation. A similar substance has been found to occur in several other rodents (Leon, Galef and Behse, 1977; Leon and Behse, 1977; Moltz and Leidahl, 1977; Moltz and Leon, 1973; Schapiro and Salas, 1970; Porter and Douane, 1977).

The diet of the mother also seems to be important in the characterization of the maternal pheromone. Young spiny mice, *Acomys cahirinus*, prefer chemical cues from females of the same species which have been reared on the same diet as their own mothers to the odour of females kept on a different diet. They also prefer cues from female laboratory mice kept on the same diet as their natural mothers to those from female spiny mice kept on an unfamiliar diet (Porter and Douane, 1977; Porter, Deni and Douane, 1977). This rather suggests that the attractive substance is not species-specific but has recognizably common characteristics across a range of species.

4.6 Growth, physical and psychosexual development

The presence of other individuals of the same species, particularly if there are many of them, has long been known to influence such processes as growth, development and sexual maturation, and ideas based on these influences remain fundamental to much of the creed and scripture of population ecology. It is often difficult to dissociate nutritional and waste product effects from odorous effects, but a few experiments have been performed with sufficient clarity for it to appear that odour involvement in growth is of widespread occurrence. Early studies with goldfish indicated that young fish grew much faster in water which contained, or had recently contained, high populations of adults. As more refined experimental techniques were applied, it became apparent that one effect of the large number of adults was to increase, through urination, defecation and small suspended particles of regurgitated food, the amount of protein in the water. But when the protein was removed, enhanced growth still occurred. Protein extracts from the skin of goldfish were found to be effective at concentrations as low as 1 part in 400 000, well below the lower level of nutritional advantage. The mucilaginous material associated with goldfish eggs also enhances growth, but only at much higher concentrations (Allee, Finkel and Hoskins, 1940). A conditioning factor resulting from the presence of crowded fish has been reported for rainbow trout and American eels, *Anguilla rostrata*. Migrating elvers are attracted to rivers containing adult eels but few other elvers. When the number of young increases, the attraction turns to repulsion. The substance, which emanates from the skin of the eels, influences growth indirectly by indicating which rivers provide suitable habitat for the species, but at the same time prevents overcrowding and subsequent diminution of resource quality (Solomon, 1977; Miles, 1968).

The odorous environment plays an important role in the growth and sexual development of laboratory mice. Most attention has been directed towards urine as the vehicle of the active ingredient, but this does not rule out the possibility of non-urine odours being at least partially effective. The pooled results of many different investigations (Table 4.5) reveal broadly that the odour of adult male mice, and particularly socially dominant mice, accelerates both growth and sexual maturation. The preputial gland secretion is not involved in this

Table 4.5 Interaction between male stimulation and female inhibition of growth and sexual maturation of young female mice. (Compiled from Cowley and Wise, 1970; McIntosh and Drickamer, 1977; Colby and Vandenbergh, 1974; and Lombardi and Vandenbergh, 1977.)

Odour donor	Neonates	
	Effect on growth	*Effect on sexual maturation*[1]
Adult male	accelerates	strongly accelerates
Adult male castrated	no effect	no effect
Adult male preputialectomized	accelerates	accelerates
Adult female virgin housed in groups	strongly retards	retards
Adult female virgin housed singly	no effect	no effect
Adult female pregnant	no effect	no effect
Adult female in late pregnancy	slightly retards	no effect
Adult female pseudo-pregnant	slightly accelerates	no effect
Adult male dominant	–	strongly accelerates[2]
Adult male subordinate	–	no effect[2]

Notes 1. Maturation revealed by first sign of vaginal introitus and vaginal connification, compared with control.

2. Maturation revealed by increase in uterine weight, compared with control.

influence, as urine removed from the bladder is as effective as that voided. Urine from adult females has no effect on growth and development if it is taken from singly housed individuals; but, if it comes from grouped females, it exerts a markedly repressive influence on growth and sexual maturation. Interestingly enough, the urethral tissue appears to be involved in this repression, for if young females are treated with the bladder urine of communally housed adult females homogenized with the urethras from singly housed females, there is no delay in their sexual maturation. In other words, the urethral tissue of grouped adult females adds a substance, or more likely a group of substances, to the urine which deactivates the maturation-delaying substance contained in the bladder urine. The urine substance has a molecular weight of 860, is positive for peptides and is apparently free of steroids and steroid derivatives. With such physical properties, it is not likely to be of high volatility, and it is not known whether it may have to be tasted as well as smelled.

The relationship between the olfactory sense and aspects of the reproductive system and behaviour in humans is, understandably, poorly defined, and any discussion of it necessitates leaving the realms of rigidly testable hypotheses. There are a few physiological observations which indicate that the relationship does exist, however. The nasal mucosa changes during pregnancy in monkeys and humans; the swelling and reddening observable may be no more than a by-product of the body's changing endocrine state, however. During menses and at puberty in both sexes of humans, nasal congestion with greatly impaired olfactory ability frequently occurs. Once again the mechanism effecting the congestion is not known. Perhaps humans are different in this respect from other animals; impaired olfactory ability associated with the heat in mammals has not been reported. For a species relying on its nose for the detection of its food or its predators such a sensory imbalance would be strongly maladaptive. In many species, such as goldfish, olfactory sensitivity increases as the females become gravid. Although his findings have recently been reexamined, Professor J. Le Magnen has shown that women have a far lower threshold to musk and musklike substances than men. Indeed, over one half of all men are anosmic to musk. Women are able to perceive the odour of exaltolide, a synthetic musk, at concentrations of 1:1 000 000 000 – one hundred times less concentrated than the lowest concentration detected by men. Their sensitivity reaches a peak at the time of ovulation and declines markedly during menses. Ovariectomized women are no different from men in their powers of detectability, but those treated with oestrogens regain their lost powers (Le Magnen, 1950). Observations such as these have been known since the nineteenth century, so it is hardly surprising that they became enmeshed in the theories of human psychology. But they did not become enmeshed in quite the way one might expect. Havelock Ellis assumed that odour was of far less importance in human sexual behaviour than in the sexual behaviour of other animals, and that personal odours of all kinds *fail to exert any attraction*, although he admits that odours can have attractive qualities if the emotional state of the receiver is suitable (Ellis, 1936). Freud ascribes to smell a very lowly role in sexual behaviour; in his view this stems from man's upright stance which isolated him from odorous sensations to be found near the earth. Arguing from an accepted close relationship between olfaction and sexual behaviour in animals, he suggests that repression of the olfactory sense, linked with the advance of civilization, has resulted in a repression of sexual activity and is at the root of sexual neuroses (Freud, 1909; Kalogerakis, 1963). Freud wrote very little about the sense of smell in humans, but other psychologists have written much and ascribed both a major and a minor importance to olfaction in the evolution of modern man. There is some evidence that the olfactory environment of young children plays an important and perhaps governing role in their psychosexual development. Sometime between the second and fifth year the onset of heterosexual reactivity ushers in the Oedipus and Electra complexes in which sexual identity is gained. Unequivocal and experimentally testable data and hypotheses cannot be obtained from interpretational studies, but clinical reports of developmental progress of young children show that the more articulate subjects frequently refer to the body odours of their parents. In one very thorough case history a young boy started to be repelled by the body odour of his father, particularly his axillary odour, at the age of $2\frac{1}{2}$, and while the repulsion never left him, at the age of 5 he could talk about it to his father. He was never repelled by his mother's body odour (Kalogerakis, 1963). This is a fascinating area for future behaviourial and psychological research, which is likely to bear important social consequences. If development of heterosexual reactivity depends upon, or is influenced by, the body odour of the parent of the opposite sex, children in one-parent families must be at some risk of impaired psychosexual development. In the clinical case mentioned above, the odour of the boy's uncle was not as repellent to him as that of his father; the odorous environment of the nuclear family is, thus, unique. The apocrine sweat glands of humans apparently have no physiological function (the sudoriferous thermoregulatory glands are eccrine) and they occur in the axillae, the perianal and inguinal regions and in the mammary areola. They commence to function at puberty and gradually senesce with age, disappearing in women at menopause. It would seem likely that their secretions have, at least some, influence on psychosexual development (Kalogerakis, 1963; Hurley and Shelley, 1960).

4.7 Summary and conclusions

Perhaps because of its great evolutionary antiquity olfaction is more heavily involved in the reproductive processes than is either vision or hearing, although these last two senses play an undeniably strong role in territorial behaviour and mate selection in a large number of vertebrates. Olfaction exerts a functional role related to a close control of every aspect of reproduction from initial attraction

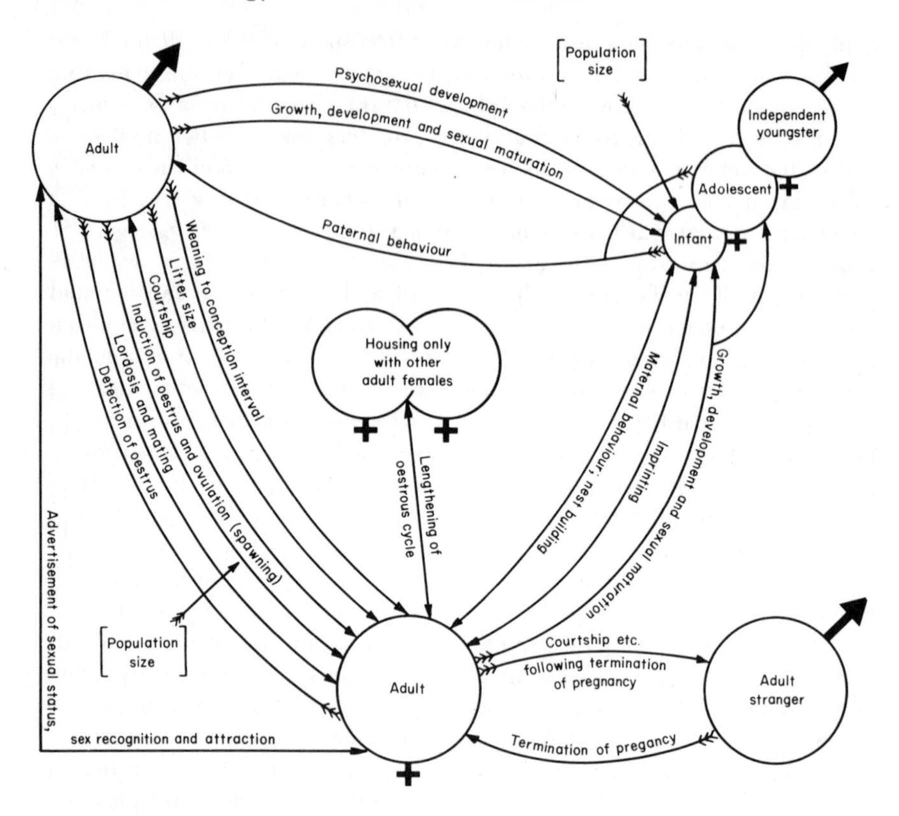

Figure 4.15 Scheme depicting the main olfactory influences in the reproductive processes of vertebrates.

and recognition of sexual status to growth rate and sexual development of the young. Close control has a high ecological desirability since mistakes made in reproduction may be energetically far more expensive than mistakes made in, say, feeding. Courtship between individuals of different species, or between male and female of differing sexual status, will be fruitless in terms of reproduction and expose both partners to risk from disruption of normal feeding and activity patterns. Such control has always carried this premium, and long ago, when organisms were unicellular and lived essentially chemical lives, control could only occur through the transference of perceptible molecules. It is likely that the olfactory dominance over vision and hearing in reproduction represents nothing more than an extension of the primeval power of transferred molecules.

Most of what is known of the interface between olfaction and reproduction is derived from studies on laboratory rodents and domestic stock and some of it seems inapplicable to naturally occurring populations anyway. The implication of this is that the generalized model shown in Fig. 4.15 must be interpreted with caution; every stage cannot be construed as having a known or demonstrable ecological effect. Predictive models of vertebrate population ecology have potentially much to gain from the olfactory model, and it is surprising that no attempts have yet been made to assess the importance of the mechanisms in reproduction in the whole pattern of population dynamics. Stated like this, the problem seems straightforward, but the reason for the reticence displayed by population ecologists is apparent on closer examination. Under field conditions how does one measure the effect of a given number of males upon the frequency and timing of ovulation in females, or their effect upon litter size? How does one assess, in numerical or energetic terms, the rôle of strange, immigrant males on the reproductive patterns of the resident female population? As the culmination and embodiment of all biological phenomena, current models in population dynamics are, at best, crude effigies of systems of unbelievable complexity.

5

Odour discrimination and species isolation

> The mechanisms that isolate one species reproductively from others are perhaps the most important set of attributes a species has, because they are, by definition, the species criteria. That there is a whole set of special devices by which the gap between species is maintained was not realized until rather recently.

With these words Professor Ernst Mayr (1963) emphasized that behavioural, as well as the previously recognized physiological criteria are involved in the maintenance of species integrity. Animals select mates according to a complicated series of signals and responses, each correct signal permitting the occurrence of a response characteristic of that species. When all the behavioural obstacles have been cleared, mating can occur and the integral nature of the species will have been preserved. Social signals, whether they be directly or indirectly involved in reproduction, utilize the sensory input channels of the receiver and are thus visual, acoustic, tactile and olfactory. Species-specific odours, like species-specific acoustic displays, are composed of many variations relating to the identity of individuals, families and populations within the species. Olfactory dialect should, therefore, appear to occur as frequently as the well-documented local dialects in bird song and mammalian vocalizations (Thielcke, 1969; Somers, 1973).

It is known that birds are able to learn the specific nuances of their particular dialect and reproduce them in their own song, although in some species, e.g., the song sparrow *Melospiza melodia*, the whole song appears to be inherited genetically (Mulligan, 1966). It is highly unlikely that vertebrate animals have any psychological ability to be able to change the quality of their olfactory displays, although the well-documented studies on the existence of olfactory imprinting (Chapter 4) reveal that young vertebrates are able to learn, and retain the mental image of, the odours of the group into which they are born. It is much more likely that differences in diet, although they may appear to be only slight, experienced by groups or populations spatially separated from one another cause subtle but important differences in olfactory display (Fig. 5.1). In the laboratory male guinea pigs are able to discriminate between the odours of conspecifics on the basis of their previous dietary history with great accuracy and to prefer that from those raised on a diet identical to that fed to the test males (Beauchamp, 1976). Such signals with characteristics which are shared among a limited group of individuals may serve several functions. They may foster the conditions necessary for group cohesion, thus serving the mutual benefit of all group members; they may also help populations which have adapted to specific ecological conditions to identify one another, and thus facilitate pair formation and mating among its individuals (Smith W. J., 1977); and, conversely, they may allow the identification of strangers. In this chapter the evidence for individual, family, population, race and species-specific odours in vertebrates will be presented, and the discussion will be closed with a consideration of odour as an isolating mechanism.

5.1 Individual odour

Probably because of their forensic application, considerable attention has been paid to the quality of individual odours of humans and their discrimination by tracker dogs. The first experiments were conducted almost a century ago through the pioneering work of J. G. Romanes who reported that his terrier was able to find him after he had headed an Indian file of 12 men across a field. Each man in the file was careful to place his feet precisely in the slots made by those in front (Romanes, 1885). In this experiment the men were genetically unrelated but studies with identical twins reveal that the genetic component in odour characteristic is very substantial. A dog given the scent of one of a pair of twins is able to track and find the other, although if the twins live in separate dwellings and experience different diets the discrimination becomes harder. Interestingly, it matters little to the dog whether it is given the scent of the palm, armpit or sole of the foot; discrimination is effected with any. The regional component, clearly discernible to humans, is secondarily acquired through the activity of specific bacteria, and added to the fundamental, genetically determined, individual odour (Kalmus, 1955).

There is little doubt that female mammals, and occasionally male mammals, have a well-developed ability to recognize their own offspring (Beach and Jaynes, 1956; Klopfer *et al.*, 1962; Nelson, 1965; Kaufman *et al.*, 1975; Kulzer, 1961), and that the same is true for those fish showing well-developed parental care behaviour (Myrberg, 1966). Although experimentation has been restricted to only a small number of species, mostly mammals, it is reasonable to assume

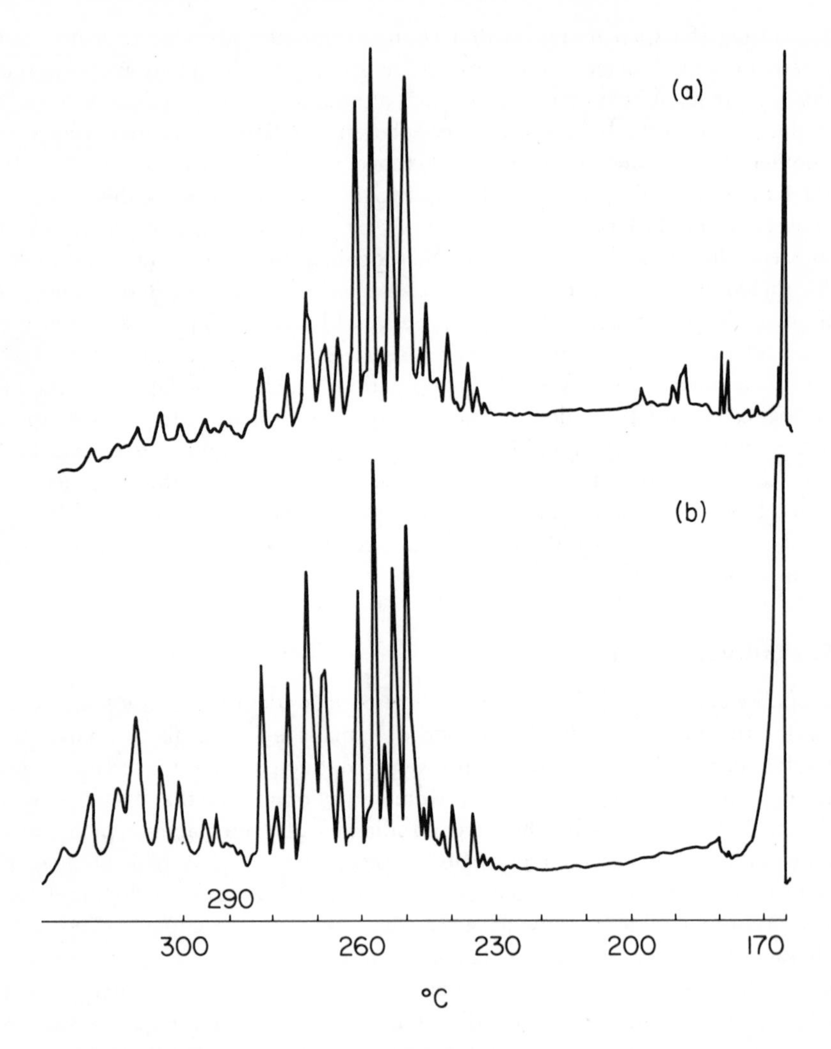

Figure 5.1 Chromatograms of sebaceous secretions from the caudal organ of the yellow-necked mouse, *Apodemus flavicollis*. Both specimens adult males, (a) from Sussex, England, (b) from Moravia, Czechoslovakia.

Mouse (a) captured 12 September 1972. Sample size 1 μl, attenuation × 1000. The sharp peaks eluting between 180°C and 200°C are due to contamination of the sample. Specimen (b) captured 4 August 1972. Sample size 1.2 μl attenuation × 1000. Temperature programme for both samples: 170°C for five minutes then 1°C per minute. Carrier gas (N_2) flow rate 54 ml/min. Column 1.5 m × 0.4 mm i.d., 3% dimethyl silicone gum on 100/120 mesh diatomite CQ support. GLC: Pye Unicam series 104, dual heated flame ionization detectors. Note the general similarity between the two samples as well as the presence of small differences. Samples (a) and (b) are wholly representative of the populations from which they are taken.

that it is a phenomenon of widespread occurrence throughout the sub-phylum Vertebrata wherever the species is macrosmatic and shows parental care. There is a strong selective advantage in a female mammal caring for and feeding only her own offspring and not those of other females and, as has already been mentioned (p. 112), foster feeding is, indeed, a rare occurrence.

The recognition of sex through odour has already been discussed. Male rats, which remain sexually active for a considerable period of time, produce in their urine and faeces a substance, or some substances, which is attractive to females. If females are kept in the presence of two cardboard tubes, both having been soiled by the same male, and then after ten minutes one tube is replaced by another tube which has been soiled by a different male, but of the same age and reproductive condition, the female shows a strong attraction to the tube bearing the novel odour (Krames, 1970). This indicates clear individual discrimination. In trials in which rodents are trained to respond to one particular odour, and for correct responses to which they receive a reward of food or water, it is apparent that individual discrimination ability is quite widespread amongst the rodents and not just restricted to intersexual discrimination (Halpin, 1974; Dagg and Windsor, 1971; Bowers and Alexander, 1967).

Individual discrimination by odour in primates has been vividly demonstrated in the ring-tailed lemur, *Lemur catta*. These primates have a thick sebaceous gland complex upon the forearm which is used for anointing the tip of the tail prior to the spectacular 'stink fights' (Jolly, 1966) in which the tainted tail is waved in the face of the enemy; under conditions of captivity some qualities of the secretions can be examined. Samples can be collected on small gauze pads held against the gland in plastic bands and then presented in quick succession to a test male. Upon first presentation, a test male exhibits much sniffing attention to a pad from another male, but further presentations of pads (which can be collected up to three months previously from the same male) elicit decreasing interest (Fig. 5.2a). If, after the presentation of four pads from one male, a pad from a different male is presented, the waning interest of the test subject is sharply revitalized (Fig. 5.2b). In these tests the pads were presented at 1.5-minute intervals; if a series of pads each from a different male is presented at the same short interval as before, a steady decline in interest occurs. But if the duration between presentations is increased to 4.5 minutes, the test male maintains a high level of interest (Fig. 5.2c and d). This suggests that discrimination between individuals can only occur after a refractory period (Mertl, 1975). Experimentation has shown the same high degree of discriminatory ability in *L. fulvus*; this species, like many primates, is typified by bold colour patterns and facial marks and olfactory discrimination of individuals is complementary to visual discrimination (Harrington, 1976; Andrew, 1963). The interaction between visual and olfactory stimuli in lemurs goes further however, for lemurs kept out of visual contact with conspecifics show little behavioural response, such as tail marking or tail flicking, to their odours. When visual contact is restored,

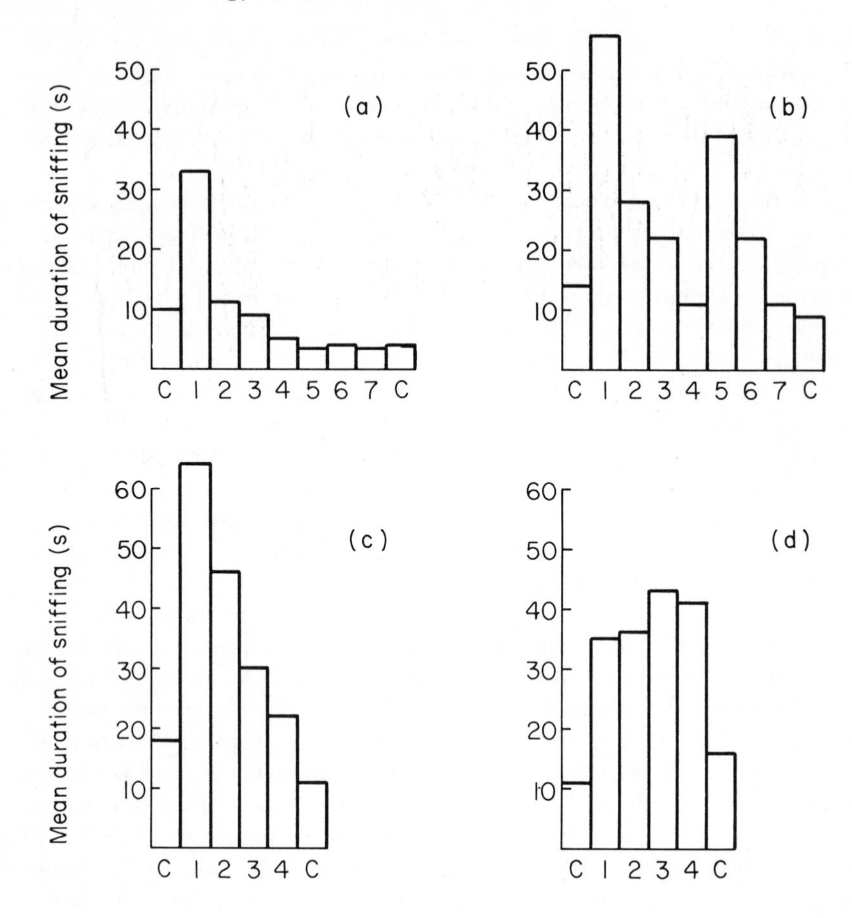

Figure 5.2 Olfactory discrimination of individual odour by *Lemur catta*. In all trials a clean control pad, c, was presented at the start and finish of testing.
(a) Seven pads, presented for 1.5 minutes each in quick succession, all from the same male.
(b) Pads 1–4 from one male; pads 5–7 from another. Other conditions as in (a).
(c) Four pads, presented for 1.5 minutes each in quick succession; each pad from a different male.
(d) Pads 1–4 each from a differet male. Pads presented at intervals of 4.5 minutes.
(After Mertl, 1975.)

however, captive lemurs show olfactory-mediated behaviour quite strongly. For example, in any conflict situation the lemur's ears lie flat against its head, and when one individual cannot smell, but can see another, the length of time the ears are kept flat against the head is very much greater than when it is in olfactory contact (Mertl, 1976).

Discrimination of individuals through scent characteristics has been demonstrated in many species, and it would seem likely that it occurs in all macrosmatic species. There is no reason to suppose that body odour is not under genetic control, and some evidence suggesting this to be so is presented below, although one must not forget that the effect of diet on the quality of odour output can be quite great. Various techniques have been employed for these demonstrations from simple choice chamber equipment (for *Mus musculus* (Bowers and Alexander, 1967) and *Rattus norvegicus* (Krames, 1970)) to the results of operant conditioning (for *Phoxinus phoxinus* (Göz, 1941), *Meriones unguiculatus* (Dagg and Windsor, 1971), *Helogale undulata* (Rasa, 1973), *Herpestes auropunctatus* (Gorman, 1976)). In very few studies has the discriminatory ability been related to an observed individual scent configuration (Gorman, 1976); this is because of the behaviourally orientated, rather than chemically orientated, nature of most experiments.

5.2 Family, population and racial odours

Families, populations and races have their own peculiar odour characteristics and response patterns, as might be expected from an extension of the genetic determination argument outlined above.

In the European water vole, *Arvicola terrestris*, analysis of the pattern of chromatographically produced peaks from flank organ secretions shows a higher degree of similarity within families than between families and within populations than between populations (Stoddart *et al.*, 1975). These observations lack behavioural experiments indicating whether odours from closely related individuals are more or less favoured than odours from more distantly related individuals, however. Studies with the salamander, *Plethodon jordani*, indicate that the reproductive state of the female governs her olfactory discrimination choice. During the non-breeding season, females show preference for the odours of neighbouring individuals, but in the breeding season a significant change in preference occurs with the females being strongly attracted to the odour of non-neighbours (Madison, 1975). Such a device may encourage breeding with unrelated individuals, since offspring tend to remain their whole lives in the close vicinity of the parent. This is certainly the case in laboratory mice. Female mice, given the choice of sawdust fouled by a brother, a non-related male of the same strain, and a male of a different strain consistently spend significantly more time investigating the odour of the same strain, but non-sibling, male. In this way the olfactory preference shown by the female is for a male whose genes are most likely to combine with hers giving the fittest offspring. There is a higher genetic risk involved in intense inbreeding (with a sibling) or in reckless outbreeding (with a male of a different strain), and the risk is contained within manageable proportions by the female selecting her mate wisely (Gilder and Slater, 1978). Local populations of the rodent *Arvicola terrestris* inbreed to a considerable extent, but emigration from and immigration into local populations occurs at a low level (Stoddart, 1970, 1971). Bank voles, *Clethrionomys glareolus*, occur over a much wider area of Britain than *A. terrestris* and the local population, or deme, concept is more difficult to apply. This is supported by the occurrence of only one subspecies of *C. glareolus* in mainland Britain – there are at least three in *A. terrestris*. In the bank vole, there is ample evidence that individuals from one race, when presented with a choice of sex partner of the same or a different race, select that of the same race by odour alone (Godfrey, 1958). This behaviour tends to reduce the chances of hybridization which has attendant biological disadvantages as will be seen later.

Because of both the clearly defined concept of race in humans and the intense interest devoted to it by anatomists, much is known about odour characteristics in humans (Baker, 1974). The Mongolid race is typified by a very weak development of apocrine sebaceous glands and the axillary apocrine complex found in Europids and Negrids is lacking. Body hair is also sparse in Mongolids, particularly in those sites where other racial groups have well-developed odour glands. Amongst the Mongolid race the subrace with the least well-developed scent glands is the Huanghoids from Korea. Correlated with the degree of development of the axillary glands is the type of wax produced by the ceruminous glands of the ears. The wax is soft and sticky in races with well-developed axillary glands but is hard and dry in those lacking underarm glands. Neither the biological nor the olfactory significance, if any, of this correlation is known. Although observations on quality of racial odour are of a subjective nature, it does appear that, as in bank voles, discrimination between races using body odour is possible. Mongolids appear almost odourless to a Europid or Negrid nose and Mongolids are very aware of the strong smell of Europeans. This appears to them to be particularly strong and unpleasant in Nordic peoples and others from northern and central Europe and indicates that considerable variations within racial odours exist. In his fascinating discussion of this topic Baker (1974) draws attention to the fact that male and female perceived odour in any one race is apparently similar, yet other physical characteristics are very dissimilar. The confusion arises because of the rancid fatty acid odours produced by bacteria living on the skin and in the hair shafts; when objective research is performed it must be carried out on pure sebaceous secretion.

Before turning to consider species odours and sexual isolation, some consideration must be given to the maintenance of aggregations, or schools, of individuals of one species. Many behavioural and physiological concepts are involved here, such as in the maintenance of breeding groups or hunting groups. All the senses are utilized to some extent; the role of olfaction in some social interrelationships is examined elsewhere in this book. Schooling, or shoaling, is a behaviour typical of many species of fish and depends for its existence upon the attractive qualities of its members. Early experiments with minnows, *Phoxinus phoxinus*, which show a strong shoaling behaviour, revealed that the site of attraction is the body mucus

(Wrede, 1932). In neurophysiological studies on the char, *Salmo alpinus*, it has been ascertained that body mucus brings about clearly demonstrable bouts of electrical activity in single cells of the olfactory bulb. A higher level of activity occurs when a test fish is presented with water containing body mucus from fish taken from the same population and home site as the subject than when the source of mucus is from a different population (Døving *et al.*, 1974). Blinding and removal of the forebrains of minnows does not interfere with their desire to form shoals – observations which suggest that if the eyes are used in group cohesion, loss of them does not necessarily result in destruction of the group. In other species the dependence of the behaviour upon the various senses differs. In the rudd, *Scardinius erythrophthalamus*, for example, blinding causes the shoal structure to disintegrate, but the individuals remain in close proximity to one another even when given the opportunity to disperse through a large aquarium (Keenleyside, 1955). When two small tanks, each containing some rudd, and one with a perforated end, are suspended in a big aquarium, the fish in the large tank gather close to the perforated tank and ignore the one with which only visual contact is possible. When fish of a different species, *Ictalurus nebulosus*, are placed in the small tanks, the rudd show no preference for either small tank or for either end of the large aquarium. This demonstrates that the attractive odour is specific to rudd and is not found in other species. In the roach, *Rutilus rutilus*, school structure is maintained by a balance between visual attraction and repulsion monitored through the lateral line (Hemmings, 1966). At night, when vision is inoperative, the complete disintegration of the group is prevented by an attraction to dissolved substances produced by the skin of the group members. This attraction serves mainly to maintain individual contact until the light intensity rises sufficiently to allow visual control once again to take over. Roach keenly follow an increasing gradient of roach odour, swimming slowly when heading up a gradient and swimming fast when away from a gradient (Hemmings, 1966). Fish schooling behaviour appears to be under the integrated control of several sensory systems, with olfaction playing a more important role at night or when test subjects have been deprived of sight.

5.3 Species odours and sexual isolation

In much of the preceding discussion is an implied acceptance of the concept that species have specific odours, in much the same way as they have other specific characteristics. Species recognition or discrimination of species odours has been demonstrated for many species of vertebrates and it is probably of universal occurrence. To list the species would be tedious and would divert attention from the main theme of indicating the role of odours in effecting sexual isolation and, hence, playing a substantial role in the process of speciation.

The bank vole, *Clethrionomys glareolus*, occurs in a series of geographical forms, or races, from many of the islands around the coast of Great Britain (Fig. 5.3).

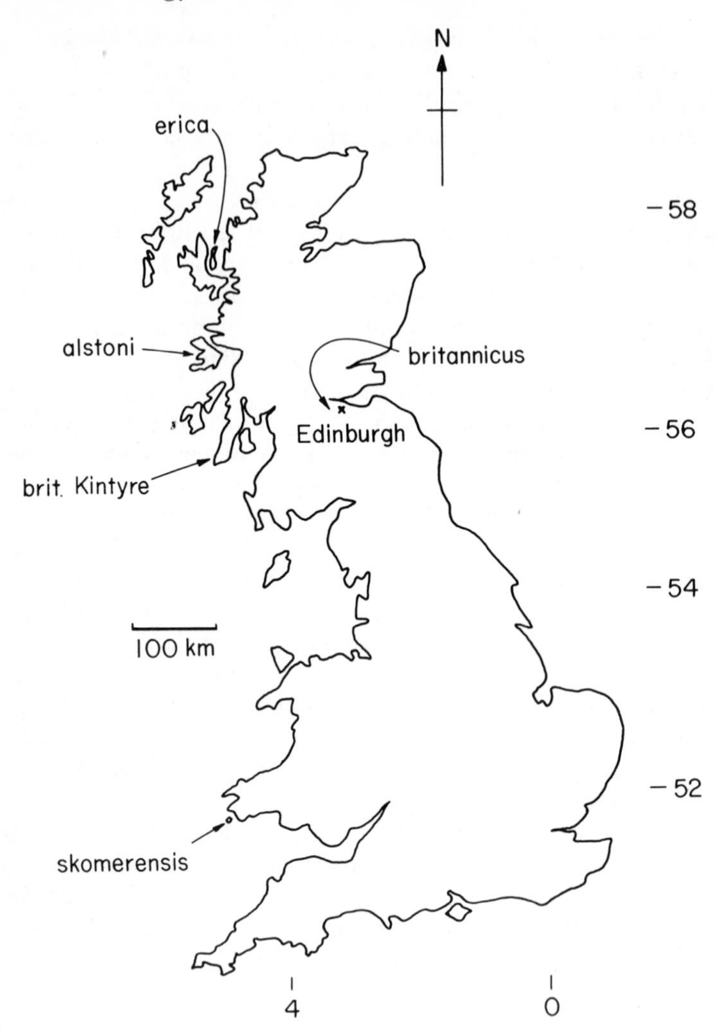

Figure 5.3 Locations of subspecies of *Clethrionomys glareolus* in Great Britain.

Crosses between forms occur without any apparent difficulty in the laboratory as long as there is no choice of mate. If a male of one race is presented with ten females, five of his own race and five of a different race, and his reproductive activity measured by a count of vaginal mating plugs forming within 72 hours, it appears that he shows considerable selectivity in favour of his own race (Table 5.1) (Godfrey, 1958). When a single male is confronted, in a choice chamber type of apparatus, with odours emanating from a female of his own race and from a female of a different race, he shows a strong attraction to the odour from the female of his own race. The significance of the attraction is shown in Table 5.2. It

Table 5.1 The mating activity of a single male *Clethrionomys glareolus* as revealed by subsequent presence of vaginal mating plugs. (Modified from Godfrey, 1958.)

Race of male	Five females of same race: number of vaginal plugs	Five females of different race or laboratory-bred hybrid	
		Race	Number of vaginal plugs
C.g. skomerensis	2	*alstoni*	0
skomerensis	5	*erica*	3
alstoni	5	*britannicus*	2
alstoni	1	*brit.* × *alstoni*	0
skomerensis	4	*brit.* × *skom.*	1
brit. Kintyre	3	*alstoni*	1

Total number of matings with same race female – 20
Total number of matings with different race female – 7

Note: Location of races as follows:
skomerensis – Isle of Skomer
alstoni – Isle of Mull
erica – Isle of Raasay
britannicus – mainland Britain. In these experiments *britannicus* specimens came from the Edinburgh region.
brit. Kintyre – Mull of Kintyre

Table 5.2 The choice of male *Clethrionomys glareolus* for females of the same and different geographical races, and for laboratory-bred hybrids. (Modified from Godfrey, 1958.)

	Male	Female selected	Female rejected	Statistical significance of choice p
	C.g. britannicus	*britannicus*	*erica*	< 0.02
	alstoni	*alstoni*	*britannicus*	< 0.01
	alstoni	*alstoni*	*skomerensis*	< 0.005
	alstoni	*alstoni*	*rutilus*	> 0.1
	skomerensis	*skomerensis*	*britannicus*	< 0.005
	skomerensis	*skomerensis*	*alstoni*	< 0.005
	rutilus	*rutilus*	*alstoni*	< 0.0005
*	*britannicus*	*britannicus*	*brit.* Kintyre	< 0.001
*	*brit.* Kintyre	*brit.* Kintyre	*britannicus*	< 0.05
†	*britannicus*	*britannicus*	*brit.* × *alstoni*	< 0.005
†	*alstoni*	*alstoni*	*brit.* × *alstoni*	< 0.001
†	*skomerensis*	*skomerensis*	*skom.* × *alstoni*	< 0.02
†	*brit.* × *alstoni*	*brit.* × *alstoni*	*alstoni*	> 0.9
†	*skom.* × *alstoni*	*skom.* × *alstoni*	*skomerensis*	> 0.2

Note: 25, 50 or 75 replicates of each trial were performed.
Locations of races as in Table 5.1.

can be seen that olfactory cues strongly influence attraction within races. Within the race *britannicus*, the geographic separation between Edinburgh and Kintyre is sufficient to cause a selective difference (see results indicated by *). Further, the table shows that laboratory-bred hybrids were discriminated against by pure races, but did not themselves discriminate significantly between hybrid and parental type (see results indicated by ⧸).

These data suggest that racial integrity is maintained, at least partly, by the response of male voles to the body odour of females. Are hybrids less viable, such that their production is selected against? Under laboratory conditions it appears that they are. F_2 hybrids live substantially shorter than F_1 hybrids and the F_1 hybrids substantially shorter than the pure races (Fig. 5.4). Since hybrids have not been recorded from the wild it is not known if they behave in a similar manner. The specific site of production of attractive odour in this species is not known, but voles of the genus *Clethrionomys* have sebaceous gland complexes at

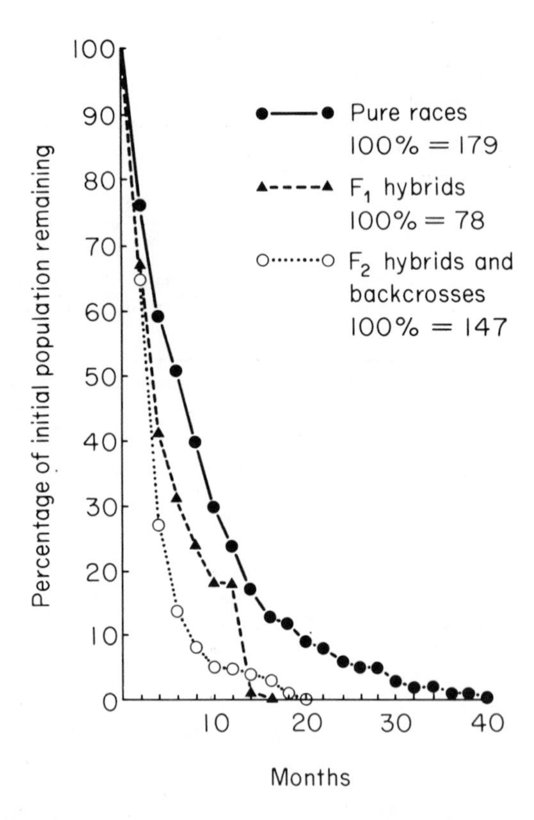

Figure 5.4 Survivorship curves, in months from birth, for laboratory populations of pure race *Clethrionomys glareolus*, F_1 hybrids, and F_2 hybrids and backcrosses. (From Godfrey, 1958.)

the angle of the mouth and in the flank-hip region. Additionally bank voles appear to use urine for purposeful marking. Voles from mainland Britain leave tiny trails of urine wherever they go while voles from the island of Skomer tend to leave a few larger drops less frequently (Johnson, 1975). Perhaps this is related to the higher degree of activity seen in mainland voles than in the island forms. Although the picture is far from clear, it is apparent that not only do odour-spreading characteristics differ between races, but odour quality appears to maintain racial integrity.

A similar situation exists for North American mice of the genus *Peromyscus*. Various species and subspecies are recognized within the genus and many species are sympatric with at least one other. Some forms, e.g., *P. polionotus* from the Florida swamplands, appear to be ecologically isolated from other species and have lost the ability to discriminate between species. Under choice chamber laboratory conditions using whole body odours of *P. polionotus* and *P. maniculatus*, test individuals of both sexes of *P. polionotus* show no consistent species-directed responses and go towards their own species odours as frequently as to the odour of other species (Moore, 1965). *P. maniculatus*, on the other hand, expresses a clear preference for its own species odour. When an oestral female of either species is used as odour donor, the responses of conspecific males are strongly heightened. The oestrous state of females appears to influence the males' choice of potential mates within this genus. Female *P. maniculatus* in dioestrus spend less time in the presence of male homospecific odour than in the presence of male heterospecific odour. But this changes markedly with the onset of oestrus. Immediately the heat starts, female *P. maniculatus* are strongly drawn to males of their own species (Doty, 1972); see Fig. 5.5. It is not known whether the attraction is real or represents an avoidance of the heterospecific odour. When mouse urine is used in a three-choice apparatus, with air as the third choice, the attractive and avoidance factors seem to be of approximately the same magnitude. Be that as it may, it seems that olfactory discrimination, in favour of maintaining species identity, occurs in these sympatric species of mouse.

The species *P. leucopus* and *P. gossypinus* are ecologically fairly well separated in areas where they are sympatric; *leucopus* selects upland and *gossypinus* lowland forests. Ecological separation is not absolute, however, for some overlap occurs in between the two forest types. Hybrids have been recorded in the wild, but rather rarely. It appears as if a sexual isolation occurs to maintain purity of line. In one series of experiments sympatric mice of the two species were taken from Leon County, Texas. Allopatric *leucopus* came from Bryan and Tillman Counties, Oklahoma, and allopatric *gossypinus* from Nacogdoches County, Texas (Fig. 5.6). All test mice were allowed to establish a nest either near to or away from an individual of the opposite sex of the two species. Mice from the area of sympatric distribution showed strong attraction to those of their own species and in most tests the choice was statistically significant. Mice from the areas of allopatric distribution, however, showed no preference for others of their own species,

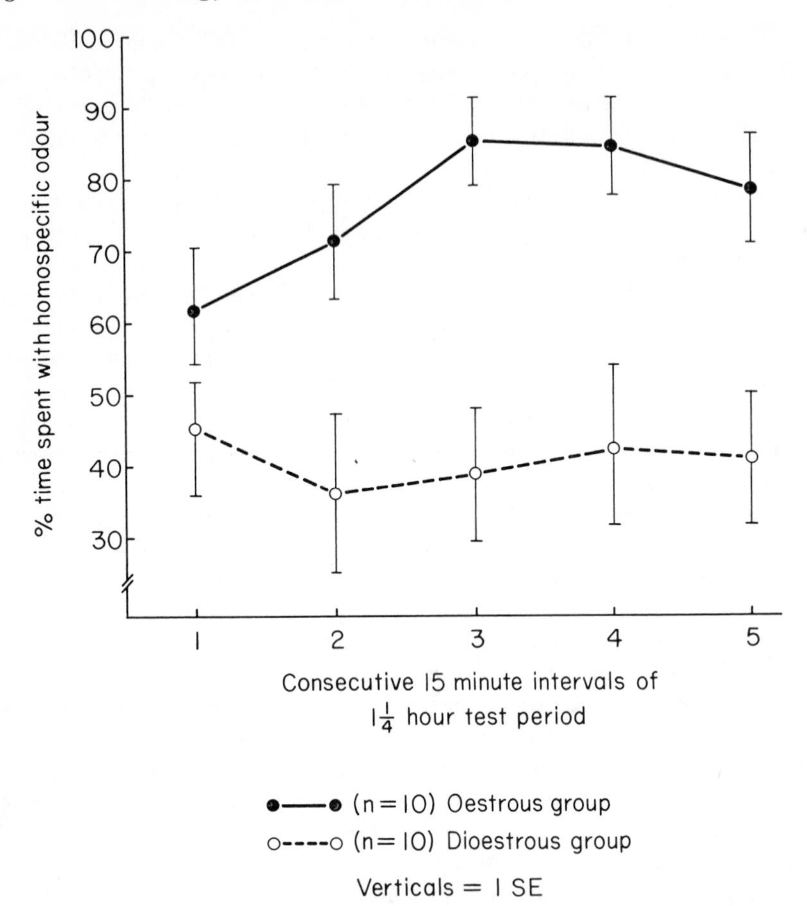

Figure 5.5 Mean percentage of time spent with homospecific male odour by oestrous and dioestrous female *Peromyscus maniculatus* as a function of consecutive 15 minute intervals in a 1.25 hour test. The heterospecific odour is *Peromyscus leucopus*. (From Doty, 1972.)

indicating that species isolation mechanisms are selected for quite strongly in regions of overlapping distribution (McCarley, 1964). This experiment did not identify odour as being either the sole or most attractive factor, but, in view of the other published observations on species isolation mechanisms in this genus, it seems likely that test subjects were guided by olfactory cues.

In Israel, the superspecies complex of the mole rat *Spalax ehrenbergi* is represented by four main chromosome forms ($2n = 52$, 54, 58 and 60), distributed clinally and paratypically (see Fig. 5.7). Hybrids of $2n = 53$ and 59 have been found near Mt Hermon in the north and in Samaria, but only rarely. It thus appears as if the clinal series represents the final stages in speciation of the

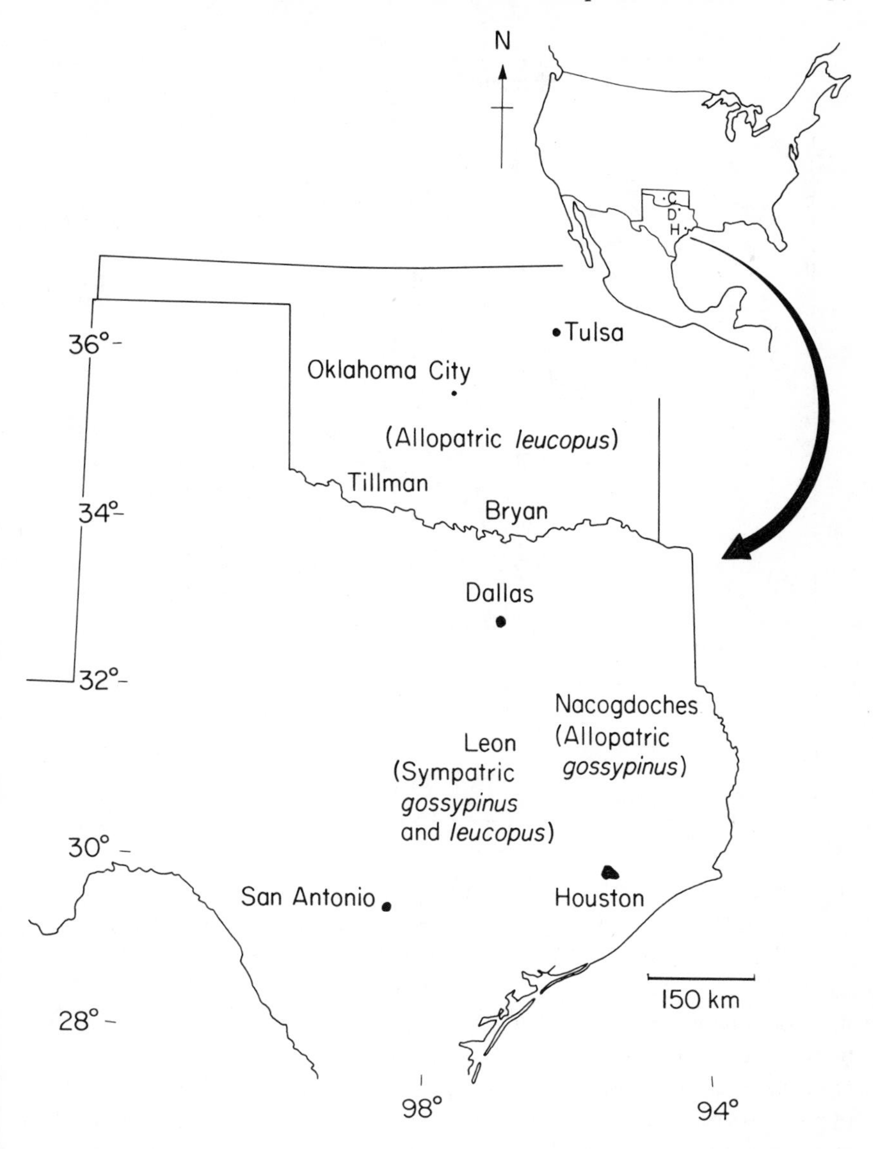

Figure 5.6 Locations of sympatric and allopatric populations of *Peromyscus leucopus* and *P. gossypinus* in Oklahoma and Texas.

four forms. It is known that the levels of agonistic and territorial behaviour are far higher in contrived laboratory encounters between heterogametic individuals than between homogametic individuals, and they are also higher between contiguous chromosome forms (2n = 52–54, or 52–58) than between

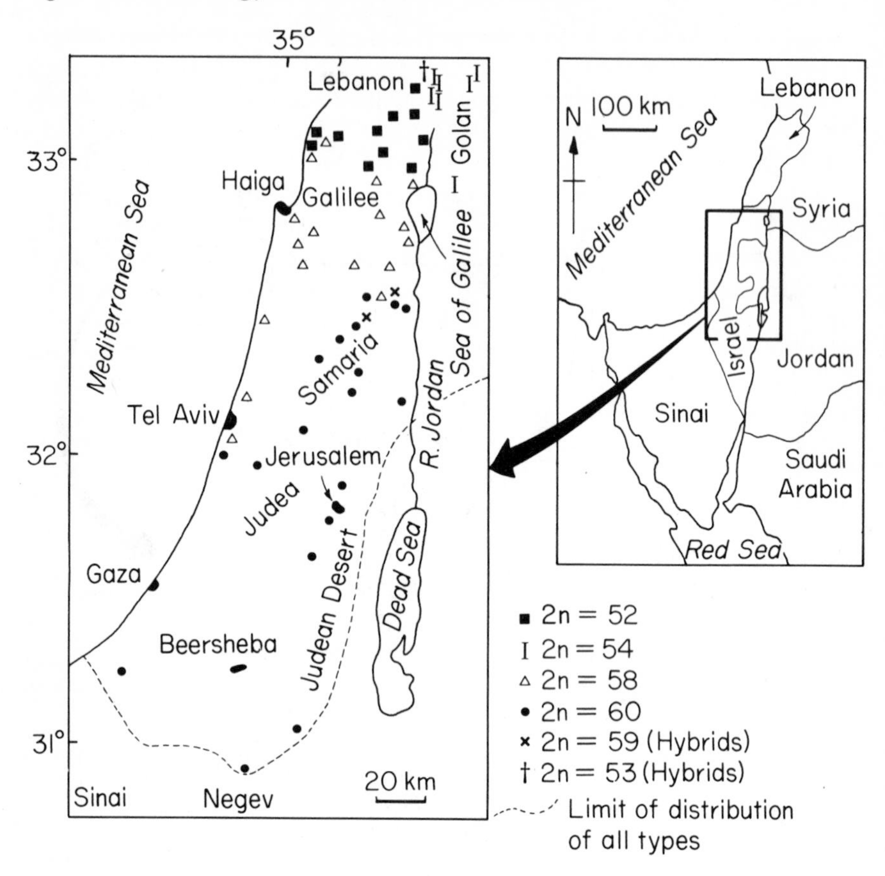

Figure 5.7 Distribution of chromosome types of the burrowing mole rat, *Spalax ehrenbergi*, in the Holy Land. (Redrawn from Nevo, Naftali and Guttman, 1975, with additions.)

non-contiguous forms (2n = 52–60). Both of these observations suggest that interspecies aggression serves as a premating isolation mechanism helping to forestall interspecies reproduction (Nevo *et al.*, 1975). If this should fail to prevent hybridization, there is another mechanism which operates at an even later point in premating behaviour. To demonstrate this experimentally, females of any particular karyotype, both in oestrus and dioestrus, were allowed to move out of a small box and enter one of two tunnels. At the end of each tunnel a small chamber held some odorous material – this was either the general body odour of a mature male represented by some soiled bedding and thus containing urine, faeces, saliva, sebum, etc., or the urine odour only of a male collected during handling and absorbed onto cotton wool. In each of a series of tests the odorous materials came from a male homochromosomal and heterochromosomal with the test female. The time spent in the proximity of both stimuli during a half-hour test is shown in Table 5.3 (Nevo *et al.*, 1976). It can be seen clearly that females in oestrus show strong attraction for the general body odour of homochromosomal males, but that dioestrus females show no such polarization of their response. The experimental data suggest that dioestrous females are selectively drawn to the heterochromosomal males, but the results are not statistically significant. The response to urine is somewhat similar and differs only to the extent that the level of significance of attraction shown by the 2n = 52 females is not as high as it is when the odorous material is soiled bedding. This may indicate that the species-specific attractant, or interspecific repellant, is at least partly present in the urine. What the whole experiment shows is that visual and acoustic cues are not necessary components of the communication environment and it is odorous cues which, at the last possible moments, prevent interbreeding.

Table 5.3 Attraction of female *Spalax ehrenbergi* of two karyotypes to general body odour and urine odour of adult males of the same karyotypes. (From Nevo, Bodmer and Heth, 1976.)

Females diploid No. and oestrus state of oestrus (*)		No. of tested females	No. of tests	Mean time (minutes) spent by female in odour receptacle				b_T (†)	p
				Homo-chromosomal		Hetero-chromosomal			
				Mean	± SE	Mean	± SE		
General body odour									
58	O	10	20	17.95	2.43	6.42	2.35	53.5	< 0.03
52	O	6	12	15.95	3.13	1.93	0.65	13.0	< 0.025
58	D	8	11	7.90	3.53	13.25	3.95	41.5	n.s.
52	D	10	12	5.33	2.27	11.03	2.45	60.0	n.s.
Urine odour									
58	O	8	18	25.32	2.32	0.63	0.50	7.0	< 0.005
52	O	8	14	8.30	3.05	5.33	2.60	27.0	0.06

Notes: (*) O = oestrus D = dioestrus
(†) b_T is the statistic derived from a one-tailed Wilcoxon's matched-pairs signed-rank test by comparing in each group the times that the tested female spent close to the homo- as compared to the heterochromosomal stimuli for every test.

5.4 Summary and conclusions

In this poorly researched field of olfactory biology, in which the most often quoted 'facts' are little more than anecdotes, a few trends appear which might, in the light of future researches, be found to have wide application.

The recognition of an individual, a population, race, subspecies and species is possible in many vertebrates by virtue of odour cues alone. That this can be shown to be so in the laboratory under often highly artificial experimental conditions does not imply that odours are any more important for discrimination than sight and sound under natural conditions. The ability possessed by vertebrates of being able to rely on one sense when one or two others are deprived has long been recognized as a fine example of evolutionary prudence.

Arguing from the above *a posteriori*, there is a growing body of evidence implicating the attractive, or perhaps repellant, quality of body odour in the processes of speciation. Species that are ecologically or geographically isolated from others appear to have lost the ability to discriminate between closely related species, an ability that is keenly developed in species not isolated in this way. The most positive responses are obtained from female vertebrates, particularly when they are sexually receptive. Other mechanisms may operate under natural conditions to keep the newly forming species apart all the time, such as is seen by the increased aggressiveness between heterogametic forms of the mole rat *Spalax ehrenbergi*; the final olfactorily induced separation functions as a fail-safe reserve system.

The relationship between the onset of reproduction and the development or recrudescence of scent organs has frequently been noted (Table 2.2; Figs. 2.1, 2.2), and a conclusion often drawn is that odorous secretions from these organs serve as sex attractants and function not only to effect intraspecific fertilization, but also, by strong implication, to prevent interspecies breeding (Hawes, 1976).

Such conclusions may be rather hastily drawn; in any case they suggest a unitary function for a great many structures which have, admittedly, a basically similar anatomical form (see Fig. 2.1). Detailed analysis of the chemical complexity of sebaceous gland secretions in rodents reveals that very fundamental differences occur between species. To take an example, there is practically no difference between the secretions of male and female *Arvicola terrestris*, if both are taken at the same time of year, but the difference between male and female *Apodemus flavicollis* is large and obvious (Stoddart, 1977). From this it is necessary to argue that the information content of the two species' secretions may be different, but no analytical studies of sufficient depth have yet been performed which might enable one to say precisely what messages are contained therein and whether a sexual isolation message is one of them. Males are socially much more active than females and a consistent feature of the entire animal kingdom is that males bear most of the adornment of the species, be it visual, acoustic or olfactory. A parsimonious interpretation of the increase in scent organ activity at the time of gonad development would be to link this activity to the social problems induced by the season of births. This aspect of olfaction is dealt with in the next chapter.

6

Dispersion and social integration

Populations of vertebrate animals are highly structured entities comprising many kinds of individual, each of which fulfils a specific social role at any given time. But the most highly structured of any found in the animal kingdom are present in the social Hymenoptera and Isoptera, and in these invertebrate groups the different social roles are performed by morphologically distinguishable castes. The crucial distinction between these and vertebrate populations is that once a caste has been determined for a particular individual, the future destiny of that individual is highly predictable, while in populations of vertebrates any one individual may, during the course of its life, perform several roles sequentially. In the extreme form of division of labour seen in the invertebrates the huge majority of colony members play a supportive role to a tiny number of fertile individuals, and never themselves become fertile, whereas all members of vertebrate populations are fertile and potentially capable of breeding. Reviewing the whole breadth of the Invertebrata, stratified populations are rare, but in the Vertebrata they are the rule. Social integration in the Hymenoptera and Isoptera is effected by the presence and continuous circulation of specific chemical substances, pheromones, throughout the colony; these always originate within the colony. Unlike the assemblages of social insects, populations of vertebrates are composed of individuals of dissimilar genetic constitution although the frequencies of certain genes may be higher in some populations than others. Although social integration in the vertebrates is governed by a series of factors extrinsic to the population, the intensity with which they impinge upon the population is a function of the population density. The factors are partly abstract in nature, and integration is effected by competition between population members for these goals. In order to be successful an individual has to assert itself over its peers, parents and offspring, and to do this it has to be *aggressive*.

Aggression is very much a feature of vertebrate biology and, although it may not be entirely lacking in the invertebrates, its occurrence is not as clearly noted. Aggression has to be directed against conspecifics in competition in such a way that neither the resource itself nor either competitor is damaged, for unsuccessful aspirants have a value, through kin selection, to their parent population. The partially abstract goals stand as surrogates for the sought-after environmental resources and become an end in themselves, the possession of which surrogates entitles the holder to the right to reproduce. There are two broad classes of goals; the class utilized by any one species depends upon the way of life of that species. Gregarious species, such as rodents and plains-dwelling herbivores, compete for a wholly abstract prize – high social status. Less gregarious and 'solitary' species compete for a physically identifiable tract of land known as a territory. Not all species fall neatly into one or other of these classes, however, and it is by no means uncommon to find species in which a group owns and defends a territory but amongst whose members there exists a strong social hierarchy. Some species showing a seasonal migration maintain a territorial system for the part of the year when the population is stationary and an hierarchical system during the migration. Scents and odours play an obvious part in these facets of integration and here, perhaps more than in other areas of olfactory biology, there is evidence that it is the *absolute* odour quality rather than learned relative odour difference which achieves the desired dispersionary end. Odours associated with dominance may be different, and consistently different, from those associated with subordinance, because of some strictly ecological advantage related to dominance, such as more or better quality food.

It is generally true that it is the males which compete for resource surrogates and hence are ultimately responsible for ensuring that the dispersion pattern of the population is suited to the quality of the environment (Wynne-Edwards, 1962). Competition and aggression are male attributes and in the overwhelming majority of vertebrate species the male sex is more highly adorned than the female. Such adornment may be visual, as in many birds and fish, acoustic, as in anuran amphibians, small passerine birds and some bats, or olfactory, as in some fish, amphibia, reptiles and a great many mammals. There are examples of both sexes bearing hypertrophied scent-producing organs (see Table 2.3), just as there are examples of both sexes carrying visual adornment (e.g., antlers in both sexes of reindeer, *Rangifer tarandus*), and it must be stated that the reasons for this are not entirely understood. Be that as it may, it is broadly true that olfactory adornment is a male characteristic, so it is perhaps not surprising to find it heavily involved in the maintenance of population dispersion. There is ample evidence derived from laboratory studies to indicate that activities normal to the social behaviour of the male are quashed if the male is rendered anosmic; carefully controlled experiments reveal that the change in behaviour is brought about entirely by the individual's inability to perceive odour, including his own (Beauchamp *et al.*, 1977; Devor and Murphy, 1973).

Animal populations live in a more or less unpredictable world in which food and other resources sometimes show marked fluctuations. When food supplies are abundant and of high quality, many members of the population may be recruited into the breeding caucus. Sometimes food supplies fail in a most spectacular fashion and when that happens the caucus shrinks to a mere nucleus. Recuritment is low and mass emigrations occur. Such a situation is nowhere better seen than by the residents of western Europe in an autumn following failure of the pine cone crop in the Eurasian boreal forests. Waxwings, crossbills and grossbeaks invade from the east in droves. Upheavals like this cannot occur unless a genetically inheritable plan for just this contingency is carried by all members of the population; the transference of such plans is of high survival value to the population. The mechanisms deciding which individuals have to leave and which can stay, as well as who is to rank where in the social hierarchy and pattern of territorial demarcation, depend upon the expression of aggressiveness by the members of the population.

6.1 Intraspecific aggression

The behaviours which follow a meeting of two adult breeding males of the same species at a feeding site, territorial boundary or traditional display place are heavily motivated by aggression. The aggressive display can be quite bizarre, involving the erection and flashing of patches of brightly coloured skin, hair or feathers, sometimes with clearly marked changes in facial expression. The very sight of one or more elements of the behaviour is often enough to release similar behaviour in others. There is, however, an increasing body of evidence indicating that certain odorous substances act as releases of intraspecific aggressive behaviour. Mice are notoriously pugnacious creatures, and good fighters waste little time in engaging in battle with all who cross their paths. If the olfactory bulbs of such mice are surgically ablated they are quite unable to show any aggression, even when they are caged with favourite sparring partners. Their response to an attack launched by others is either to flee or to adopt a submissive posture. This suggests that the olfactory bulbs act not only as the seat of a releasing mechanism, but also as the seat of an arousal mechanism; otherwise they would have reacted to being bitten. That it is the odour of the mouse which releases an aggressive attack by an opponent can be demonstrated by completely enveloping one of a pair of fighters in an atomized mist of commercially available perfume. Such a treatment increases the latency of attack by about 3½ times and the mean number of attacks falls by a factor of three. Clearly, in matters of honour, male mice are led by the nose (Ropartz, 1968).

Where is the site of production of this aggressive promoting substance? Mouse urine has been shown to be involved in many aspects of physiology and behaviour, particularly in reproduction, and it can also be shown to be a powerful aggression releaser. Because both sexes of mice, and non-aggressive as well as aggressive males, require to remove dissolved metabolic wastes from the body, it is unlikely that any specific substance associated with aggression can be produced by the kidneys, ureters or bladder. But mice, like all mammals, possess paired preputial glands which lie at the base of the urethra and open into it. Using the latency to the first bite, the number of bites delivered and the accumulated attacking time as a measure of the effect of the sebaceous preputial gland extract taken from adult males and daubed onto one of a pair of trained fighters, the potency of the extract in increasing the level of overall aggression is striking (Fig. 6.1) (Jones and Nowell, 1973). Female mice have poorly developed preputial glands, and in juvenile males the glands develop only at

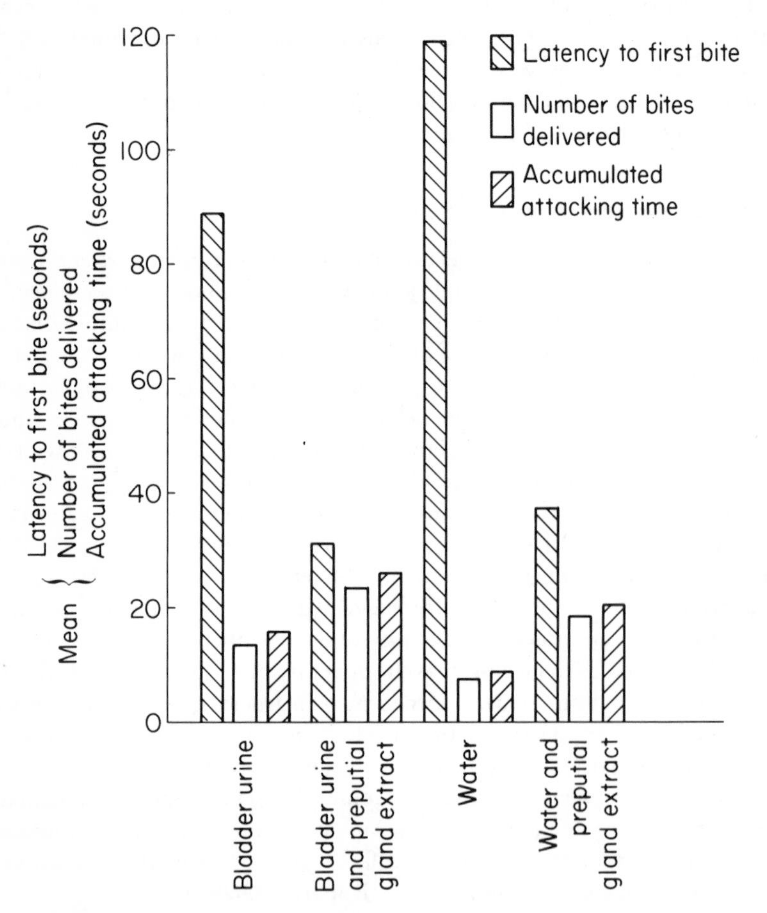

Figure 6.1 The effect of preputial gland extract on certain measurements of aggressive behaviour in the laboratory mouse. (From Jones and Nowell, 1973.)

puberty. Both these classes of mice suffer far fewer attacks from adult males than other adult males, but it appears there is more than one mechanism operating here to afford protection from attack (see Chapter 7.4) (Dixon and Mackintosh, 1971, 1976). The urine of adult and juvenile female mice, and the urine of ovariectomized adult females, possesses an aggression-inhibiting quality. The active ingredient in adult females is present in the bladder urine so is not the product of preputial or vaginal wall glands (Evans *et al.*, 1978). (It is interesting to note in passing that the sex attractant quality of the urine of female mice is removed following ovariectomy; ie., removal of the site of production of sex steroids.) The urine of young males, in contrast, fails to inhibit the aggressive behaviour of an adult male when daubed onto the back of another adult male, yet when a juvenile male is introduced into an enclosure containing a territorial male it is protected from attack. A substance, or series of substances, produced by the plantar glands of the feet has been postulated (Ropartz, 1966) but not confirmed by experimentation. The absence of blood testosterone, prior to puberty, leaves the preputial gland in a quiescent state and it is the lack of an aggression-promoting substance, rather than the acquisition of a protective substance, which accounts for a life free from harassment enjoyed by juvenile males. Visual and acoustic cues undoubtedly also aid in this protection (Sewell, 1967). Adolescence brings about significant changes in the production of secretions by specialized sebaceous glands in other species of mice, and it has been suggested that the sudden acquisition of adult odour characteristics initiates dispersal from the family nest (Stoddart, 1973, 1976).

Not all additives to the urine of adult male mice enhance aggression. The coagulating glands produce a series of substances which enter the urine in the urethra and which decrease significantly the aggressiveness of trained fighters (Jones and Nowell, 1973). Why mice should manufacture simultaneously two substances which work in opposition to one another is a bit puzzling, but the interactive mechanism of the two may allow for careful fine-tuning of aggressive behaviour by the neuroendocrine system.

Perception of an alien odour not infrequently gives rise to an expression of threat behaviour in which the perceiver goes through a behavioural repertoire not normally seen in the absence of another animal. This does not mean that threatening behaviour is initiated only by odour; it means that in the context of the state of awareness shown by a would-be infiltrator the odour signal emanating from a scent mark is capable of releasing the behaviour. Beavers, *Castor canadensis*, 'hiss' as they deposit castoreum on scent mounds belonging to alien colonies. Such behaviour is normally seen in physical encounters and is characteristic of aggression and conflict (Aleksiuk, 1968). Hamsters are observed to grind their teeth while puffing up their cheeks when smelling the scent from another (Eibl-Eibesfeldt, 1953), and similar expressions of unease are seen in a host of other species. Even domesticated male dogs are prone to growl and show dorsal piloerection when confronted with some alien urine marks. Many, indeed the majority, of scented lamp posts elicit little more than intense sniffing; only a few, presumably those tainted by dogs of about equal dominance, induce overt threatening behaviour.

6.2 The social hierarchy

Although the role of olfaction in maintaining the social hierarchy has been clearly demonstrated, it is still not known whether a certain odour quality is the cause or the effect of the ranking. A complicating factor is that odorous secretions frequently fulfil more than one function, so it is sometimes difficult to relate an olfactory observation to a social observation with a high degree of certainty. An example of this is seen in the rabbit, *Oryctolagus cuniculus*. Both sexes of rabbits bear chin glands, specialized submandibular glands opening onto the underside of the chin. Until they are about 50 days old the chin glands of males and females are more or less the same size. Puberty sets in at this time, and immediately the glands of males start to increase enormously until those of the oldest males weigh twice as much as those of the oldest females. Two factors influence the actual size of the males' glands and they are neither always nor necessarily related. Firstly, socially dominant bucks have significantly heavier chin glands than socially subordinate bucks, and exhibit 'chinning', or scent-marking behaviour, many times more frequently. The same trend, although at a much lower rate, is seen for females (Fig. 6.2b). Secondly, the size of the chin, anal and inguinal glands of

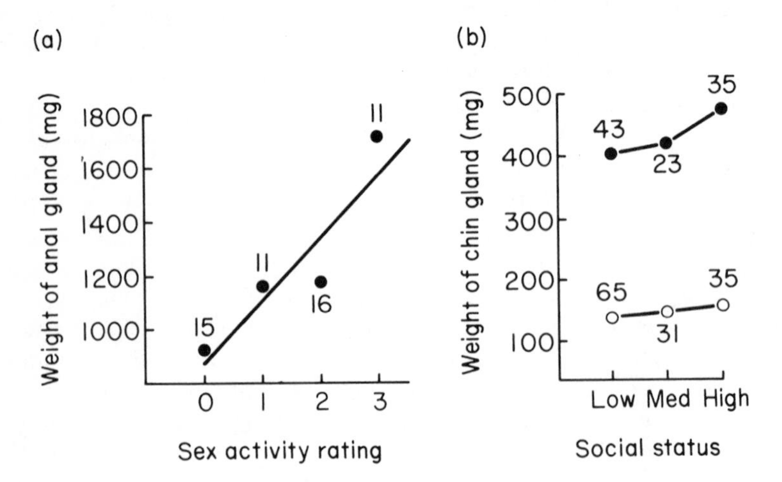

Figure 6.2
(a) Relationship between weight of anal glands and sex activity of male rabbits, *Oryctolagus cuniculus*. (o = inactive; 3 = very active);
(b) Relationship between weight of chin glands and social status of rabbits. (●–● males; o–o females).
Small numerals denote sample size. (From Mykytowycz and Dudzinski, 1966.)

bucks is strongly related to the level of sexual activity shown. The anal glands of sexually active bucks may weigh half as much again as those of sexually quiescent bucks (Fig. 6.2a). Chin, anal and inguinal gland size is strongly influenced by blood testosterone, but the exact relationship between sex hormone level and sexual activity rating is unclear (Mykytowycz and Dudzinski, 1966). In other groups the relationship between gland size and dominance status is less confusing. For example, many of the New World platyrrhine primates bear sternal sebaceous glands the size of which is directly related to dominance level (Epple and Lorenz, 1967).

The evidence indicating that different odour qualities are associated with different states of dominance is rather easier to assess. Rabbits use their veritable battery of scent glands for quite distinct purposes. The chin glands are used for scent-marking objects at territorial boundaries, but the anal glands are used for tainting faecal pellets which are deposited at dunghills. Dunghills are sited on raised areas close to the warrens and serve only the males of the community; females seldom come to them and juvenile males are uninterested in them (Mykytowycz and Gambale, 1969; Mykytowycz and Hesterman, 1970). Using a group of human testers to assess odour strength, Mykytowycz and his collaborators showed that faecal pellets from these sites have a more intense odour than non-marking pellets. Anal gland secretions from dominant bucks have a more intense odour than secretions from subordinates. Male anal gland secretion always smells more intensely than female anal gland secretion, although increasing age in both sexes is accompanied by increased intensity of odour. As far as the human testers are concerned, the odours achieve their strongest intensity during the breeding season (Hesterman and Mykytowycz, 1968).

This type of analysis tells little about the actual manner in which the odour signal works; like traffic lights, odour signals can halt progress, warn a traveller to take great care, or they can encourage further progress. Observations at dunghills reveal that strange rabbits are not inhibited by the odour of the marking pellets; on the contrary, adults and especially dominant males are attracted to these places. By watching the subsequent progress of an intruder it appears as if the signals it receives from a dunghill indicate to it that it is entering an unfamiliar sphere of interest and is subject to aggressive treatment (Mykytowycz and Hesterman, 1970; Mykytowycz and Gambale, 1969). Studies with rats show that the traffic light stands at red for them. Rats from stable hierarchies spend significantly longer in the vicinity of a cardboard cylinder which has held a submissive strange male than in the vicinity of an identical tube which has held a dominant male. In other words, the odour of dominance acts to release withdrawal behaviour on the part of other males. Strangely enough, it seems that this response is only developed in rats which come from a stable hierarchy; rats from temporary social groups respond about equally to both odours, but such temporary groups may be rare under natural conditions (Krames, Carr and Bergman, 1969).

In these examples the signals containing the dominance message are left behind by the animal expressing that dominance. Frequently, however, it appears that the scent-marking behaviour itself is as important as the deposited secretion, and ethologists have amassed considerable data on the frequencies of scent-marking behaviours displayed under a variety of conditions. The general rule is that males not only mark more frequently than females, and that dominants mark more frequently than subordinates, but very often dominants mark most when in the presence of their subordinate rivals. This is seen in a great many mammalian species and is summarized in Table 6.1 for a few species; it may be taken as an example of a widespread trend throughout the mammals.

Living in an aquatic environment, fishes are not able to deposit scent upon either conspecifics or objects. Bullheads, *Ictalurus natalis*, are highly social and show a strong social hierarchy. If a fish is removed from a tank in which it was dominant to another and replaced after a short isolation, the resident submissive fish accepts the returned fish as a dominant and does not attack. But if, during the period of isolation, it experienced a losing encounter with an even more dominant bullhead, the originally submissive one immediately attacks it. It appears that these mildly stressful situations are sufficient to cause a noticeable change in some aspect of body chemistry, but whether this affects the mucus covering the body or the urine is not known (Todd, Atema and Bardach, 1967). Presumably the odour diffuses outwards through the water from the fish but carries much the same information as the odour diffusing from a deposited scent mark.

Little is known about odour in the expression of social dominance in amphibians, reptiles and birds. It is likely to be of some importance in reptiles since in this Class of vertebrate are seen some of the most complex olfactory interrelationships outside the mammals. There is no evidence of it occurring in any species of birds.

6.3 Territoriality

Animals have evloved many means of most optimally dispersing their populations throughout the available suitable habitat. Although almost all species can be regarded as defenders of a small exclusive space around them, some demonstrably defend a very large stretch of habitat in which they spend the majority of their time. Such are the truly territorial species. The term 'territory' covers a multitude of functions, for in some species the territory is for mate attraction, and in others it is for the provision of adequate food for the rearing of young. The spacing out of the individual members of a population may more successfully combat predators than the population living in large groups. We need not be detained by a consideration of functions; this is adequately dealt with in most modern ecological texts (e.g., Krebs and Davies, 1978). What concerns us here is the role played by scent in the maintenance of territories. There is much

Table 6.1 Characteristics of scent-marking behaviour associated with dominance in a representative sample of mammalian species.

Species	Main characteristics associated with dominance	Source
Mesocricetus auratus	Dominant males work at higher frequency than subordinates. This occurs even when dominant and subordinate are not housed together. Flank marking at high intensity is correlated with agonistic tension. Negatively correlated with tendency to flee. Dominants mark most in subordinates' home boxes. Subordinates mark most in or near their nests and this action appears to offer some defence against intruding dominants.	Johnson, R. E., 1975
Meriones unguiculatus	Dominant males scent mark (ventral rub) more than any other dominance/sex class when in an open arena.	Thiessen, 1968
Mus musculus	Prior to pairing, adult male mice urinate copiously and freely throughout the environment. After pairing and a fight for the establishment of dominance, only the dominant urinates throughout the environment, the subordinate urinating only in one or two restricted places.	Desjardin, Maruniak and Bronson, 1973
Oryctolagus cuniculus	Dominants of both sexes chin mark up to eight times more frequently than subordinates, but males mark more frequently than females.	Mykytowycz, 1965
Callithrix jacchus	Dominant males mark with their circumgenital glands more frequently than subordinates; dominant females mark more frequently than subordinate females. Only the highest-ranking males mark with the sternal gland, and then infrequently.	Epple, 1974
Cephalophus maxwelli	In large arena dominant males use suborbital gland more frequently than subordinates. All males mark at least twice as frequently as they do when housed alone when they are housed in a group. Most females mark much more frequently when in the presence of other females. Sometimes they mark each other.	Ralls, 1971
Muntiacus reevesi	Males mark more than females and dominants of both sexes mark more frequently than subordinates. Some attempt to cover marks left by others.	Barrette, 1977
Dicotyles tajacu	Marking of conspecifics with the dorsal gland restricted only to the dominant in any pair. Subordinates never observed to mark.	Sowls, 1974

evidence indicating that would-be intruders perceive at its boundary whether a territory is occupied or not, and if it is they more often back away or cross it at agreed places than venture further. This does not mean, however, that territorial boundaries are inviolable barriers; neither does it mean that an intruder risks certain attack if it ignores the peripheral messages. Territorial defence may consist of three or more skins of information broadcasting, the final one of which occurs when the resident physically chases the intruder out. The front line may be strategically positioned odour cues and the second a visual threat display. Whether or not a particular intruder penetrates one, two or three lines of defence may depend upon its own physiological state and the information it receives from the front line.

Territorial scent-marking is a well known and often reported characteristic of mammalian behaviour, but it is rather little known in other classes of vertebrates. Possibly it only occurs in the lizards, but behavioural confirmation is sparse. Comparison of scent gland structure and cyclicity of secretion production of lizard glands with those of mammals reveals many similarities of detail. In both animal types the glands hypertrophy at the onset of sexual maturity or at the start of breeding and atrophy at the close of breeding or resulting from castration. Territorial behaviour is well marked in lizards and it seems quite likely that the secretions of these glands are used as the front line of territorial demarcation (Gasc, Lageron and Schlumberger, 1970; Cole, 1966).

There is little known about the role of olfaction in territorial behaviour of birds, but it is probably quite uncommon. The piscivorous, colonially nesting sea birds, many of which have good olfactory powers, may claim nest-burrow ownership and retain possession using odour cues. The source of the odour is the strongly smelling stomach oil in such species as Leaches petrel, *Oceanodroma leucorrhoa*, and Wilson's petrel, *Oceanites oceanicus*, which is expelled during territorial disputes and may build up to a depth of 20 millimetres around the burrow entrances (Stager, 1967). Although no experimental proof of territory marking in this way is available, the known olfactory ability of these birds is quite adequate to render such an interpretation perfectly feasible.

Amongst mammals, scent-marking behaviour is known to occur in almost every terrestrial order (Thiessen and Rice, 1976), although its involvement in territorial maintenance is known for only a few species. In some species the same behaviour is seen in two contexts each with quite different motivations. For example, dwarf mongooses, *Helogale undulata*, scent-mark their territory in precisely the same manner as that in which a male scent-marks a female newly in oestrus (Rasa, 1973). Scent marking is not restricted to territorial species, and occurs in gregarious species which exhibit a clear social hierarchy. This is the case in house mice, and the information contained in the urine mark appears to convey information relating to the depositor's social status rather than to restrict the access to a particular piece of habitat.

One of the chief reasons for ascribing a territorial demarcation function to many scent-marking behaviours is that there is a tendency for marks to be applied more frequently in those zones where individuals meet than in those parts which are more exclusive. This is by no means a rule and many species pepper their entire territory with a blanket coverage of scent marks (Fig. 6.3). Most critical observations have been made on mammalian carnivores, probably on account of their being strongly territorial and possessing clear scent-marking behaviours, but a rather more complicated picture emerges than might have been expected.

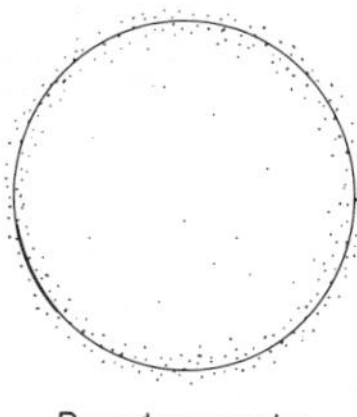

Boundary marks		Hinterland marks	
Vulpes vulpes	*Canis aureus*	*Herpestes auropunctatus*	*Lutra lutra* (marine habitat)
Canis familiaris	*Felis tigris*	*Hyaena brunnea*	*Diceros bicornis*
Canis lupus (pack)	*Vombatus ursinus*	most ungulates	*Oryctolagus cuniculus*
Meles meles	*Galago alleni*	most primates	
Lutra lutra (riverine habitat)	domestic cat		
Crocuta crocuta			

NB: In both categories there may be a concentration of marks in the immediate environs of the home site.

Figure 6.3 List of mammals for which there is clear evidence of territory marking, arranged according to whether the marks are principally placed either at the territory boundary or scattered throughout the hinterland. In both categories there may be a concentration of marks in the immediate environs of the home site.

It appears that the environmental conditions exert a strong influence on territorial demarcation, and this is particularly well seen in northern temperate species which must, in order to survive, be catholic feeders. In predictable environments a rather constant pattern of territoriality occurs and with it a constant demarcation procedure. Spotted hyaenas, *Crocuta crocuta*, defend clan or group territories in open grassland, savanna bush and shrubland, and regularly patrol the boundary zone between their own and a neighbouring clan's territory. As they go they deposit their faeces in huge piles right on the boundary. Apart from the smell of these middens, they serve also as a visual mark, and are usually positioned near to tracks or other already existent landmarks (Kruuk, 1972). Such a pattern appears to be constant. In the European otter, *Lutra lutra*, on the other hand, the pattern of demarcation is heavily influenced by the specific environment of the animals. Where otters occupy a riverine habitat and establish a holt in a bank within the sausage-shaped territory, scent-marking sites are concentrated at the territory boundaries. Otters mark by defaecating and at the same time apply a small amount of gelatinous anal gland secretion to the faeces. This behaviour is called 'sprainting', and spraint sites are used by all the animals sharing a boundary. Thirty to forty such sites may be found in each kilometre of overlap area, and a single dog otter may have to patrol three or four kilometres of boundary (Erlinge, 1968). The otter is as much at home in the sea as it is in rivers, and in the heavily fjorded parts of northern Scotland otters establish holts close to the shore line and hunt off shore, in and around the archipelago of small islets. Here the distribution of spraints is quite different, with a great concentration within 50 metres of the holt. This difference in pattern, which is summarized in Fig. 6.4, can be explained by the major habitat differences relating to food availability. In riverine habitats an intruder can enter an occupied territory only by crossing a territorial boundary. It is therefore desirable to keep the boundaries clearly demarcated. Marine otters, however, can approach any holt from the sea without encountering any sprainting site, so under these conditions it is advantageous to scent heavily the holt's immediate environs so that an otter can land anywhere and immediately assess its proximity to the centre of an occupied stretch (Kruuk and Hewson, 1978).

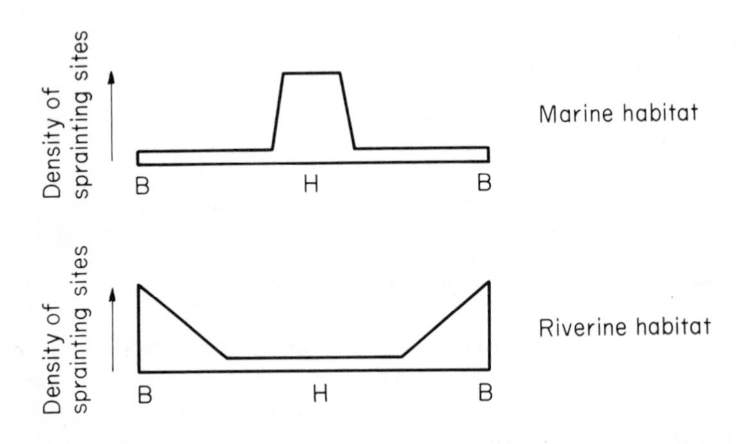

Figure 6.4 Diagrammatic representation of density of sprainting sites in territories of otters, *Lutra lutra*, living in marine and riverine habitats. H. holt; B. territorial boundary.

Field studies on other terrestrial carnivores have tended to centre on one habitat type, and there are not as many comparable data available on them as there are for the otter. Both the fox, *Vulpes vulpes*, in woodland Britain and the wolf, *Canis lupus*, in subarctic North America urinate much more frequently at the territory boundaries than in the hinterland. Foxes place a token mark of a few drops of urine at hundreds of localities within their territories and are strongly inhibited from urinating when inside a neighbour's territory (Macdonald, in press). Along well-traversed paths in the territory there may be more than one urination site per metre of track; though not every site is used on every occasion the resident checks his range. If an alien site is experimentally transferred into an established resident's territory, it is immediately investigated excitedly and overmarked. However, if the resident fox is taken into the territory of the individual whose mark it had obliterated in the experimental transference trial, it never once attempts to repeat the overmarking. It appears that the responses of an animal to territorial demarcation signals depend upon a complex interaction of factors, even though the perceived information must be the same whether the site from which it comes lies within or without the perceiver's own territory.

Wolves live either singly or in pairs or packs, and it is more usual for them to live in smaller aggregations during the season of high food abundance. Wolf packs urinate slightly more than twice each kilometre and over 50% of all urinations and defaecations occur at the territorial boundary (Peters and Mech, 1975). Single wolves, however, lay most faeces deep within the forests and actively avoid placing them in overlap zones. Pairing, however, seems to work wonders to wolves' morale and immediately there is an expression of territorial scent-marking with faeces being placed conspicuously on paths, tree trunks and other prominent places.

Figure 6.5 Badger scent-marking. (Photograph: H. Kruuk.)

Badgers, *Meles meles*, in English woodlands live in groups called 'clans' and each clan occupies a territory. The territories are marked by the badgers digging a little scrape and dropping in some faeces. Sometimes dark yellow jelly-like matter from the anal pouch is placed on top of the faeces, and occasionally only the gelatinous material is deposited in the scrape. This material is applied during a rapid squatting behaviour during which the whole perineal region is rubbed along the ground (Fig. 6.5). Almost 70% of all latrines, as the scrapes are called, are sited within 50 m of the territory boundary, while only 11% occur within 50 m of the sett. This relationship is shown in Fig. 6.6. Furthermore, the number of defaecations comprising each latrine is almost three times as high at the boundary (average 14.3) as it is close to the sett (average 5.4) (Kruuk, 1978). This ratio between marking frequency at the boundary and in the hinterland is seen in several species; for example, in the bushbaby, *Galago alleni*, urine marks are applied three times more frequently in the territory overlap zone as in the centre (Charles-Dominique, 1977).

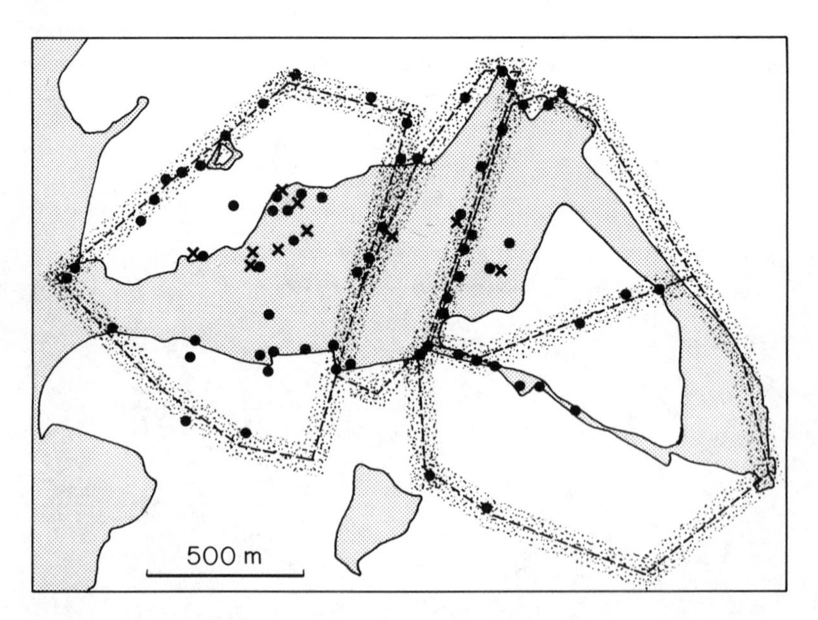

Figure 6.6 The relationship between latrine sites and range boundaries in four clans of badgers, *Meles meles*. The boundary positions were determined by radio tracking. Boundary area = 50 metres either side of the boundary. •; latrines; x; setts; Stippled area; woodland. (After Kruuk, 1978.)

A distinctive and universal feature of mammalian scent marks is that they are always made as obvious as possible by being deposited high up on a structure such as a rock (Fig. 6.7b) or a tree trunk, or by being positioned close to an

Figure 6.7
(a) Otter, *Lutra lutra*, sprainting site, on a grassy bank close to the shore of a lake.
(b) Otter faeces positioned high on a rock at the edge of a lake.
(c) Otter scrape in soft sand.
(d) Part of an otter slide. Scent from the ventral part of the body is deposited on the snow covering of an otter's path by this activity. (Photographs: S. Erlinge. (a) – (c) Reproduced by kind permission of *Oikos*.)

obvious landmark, or by being made startlingly contrasting with the environment in some way, perhaps by the scratching up of the soil (Fig. 6.7c), or the digging of pits. These are not exclusive alternatives and animals often utilize one, two or all of them. It is not unusual for marking sites to be used for many years, and when this happens physical and nutritional factors influence the surrounding vegetation. Otters use sprainting sites for decades and the effect on the surrounding vegetation is quite remarkable. Herbaceous plants react to the high levels of nitrogen by giving way to nitrophilous species of grasses, such as *Holcus lanatus*, and lichens and algae (Fig. 6.7a). The grasses react to the high nutrient levels and in spring and early summer colonize the decomposing faeces; gradually the sprainting site grows and can protrude 15 cm or more above the surrounding grasses. Almost as remarkable are the middens produced over the years by spotted hyaena packs. A large communal defaecation site can cover up to 0.25 hectares and contain the decomposing remains of thousands of faeces. Because of the high bone content of hyaena diet, the faeces are pale in colour and sometimes pure white. The purest white faeces were for long collected by the ancients as a source of the white pigment called *Album Graeca*. Hyaena middens are visually very striking, and the message they broadcast cannot be ignored. Although these sites themselves indicate territory occupation, the fact that the residents have to come to them for defaecation and urination may itself have an important signalling function.

Natural landmarks help emphasize the presence of a scent-marking site. Spotted hyaena middens tend to be positioned close to a track intersection or a pile of boulders, so it is easily perceptible by travellers passing along the track. Territorial animals are strongly attracted by vertical objects; domestic dogs may pull their handler across a road to visit a lamp post, and domestic cats tend to bury their faeces close to a wall or a bush. Badgers deposit their faeces significantly closer to landmarks, e. g., a car track or the woodland edge, than would be expected by chance (Table 6.2). It is clear that the visual and olfactory components of scent-marking sites are closely integrated.

Table 6.2 Analysis of proximity of badger, *Meles meles*, latrines to landmarks. In all cases landmarks are located within 10 metres, and usually 3 metres, of the latrine. The observed dispersion is compared with an expected dispersion assuming random deposition of faeces in the same area and their location near landmarks. (Modified from Kruuk, 1978.)

Type of landmark	Observed No. of latrines $n = 111$	Observed %	Expected %
Car track	44	39.6	11
Vegetation boundary (e.g., woodland edge)	38	34.2	14
Fence (usually barbed wire)	35	31.5	15
Conifer	10	9.0	0
Other (pylon; tree in field)	2	1.8	2
None	35	31.5	69

Most dog owners will be mildly aware that there is sometimes an acoustic component to scent-marking; a particularly rich lamp post may elicit quite considerable growling while the dog is adding its urine to that of others. While this noise could be motivated by the information received from the post, it is likely that, in some cases at least, it is intended to draw attention to the marker. A loud and throaty hissing is normally an accompaniment to the deposition of castoreum by beavers, *Castor canadensis*, on their scent mounds. The sound travels over quite great distances and combines with the sight of the mound and the musky odour of the secretion to advertise territory occupancy (Aleksiuk, 1968). It is quite necessary for beavers to control their population dispersion very precisely since their food, primarily stands of white aspen, is extremely vulnerable to overexploitation.

If territories are not absolutely exclusive and their boundaries not inviolate, what advantages accrue to a territory holder? For many carnivorous species the territory fulfils its primary purpose of providing its tenant with an adequate supply of food for its needs and that of its family. But rabbits are strongly territorial and show much scent-marking behaviour, yet their food is in great abundance and will only be limited under the most extreme cases of ecological imbalance. What information do the copious scent marks convey to the resident animal and intruder? Intruding rabbits are strongly attracted to latrine sites and clearly go out of their way to sniff them. Just like the displaced fox, the intruding rabbit makes not the slightest attempt to contribute to the dunghill – it merely extracts information from it. What this information is we do not know, nor do we know what effect it has on the intruder. The real benefit of the territorial system is felt by the residents in occupation. Under natural conditions both male and female rabbits who are on their own territory, and therefore surrounded by their own odour and that of their clan, win two-thirds of all aggressive encounters. Male rabbits gain most confidence from the odour of their own chin and anal glands, and a little confidence from the smell of their own urine. Their own inguinal gland secretions have no effect. Females gain confidence in encounters with other females and win more fights when surrounded by their own urine and anal gland secretions, but inguinal gland secretion odour is of little use. For both sexes the odour of a familiar cage mate, of the same or opposite sex, significantly increases their chances of winning. Chin and anal gland secretions, along with urine, are used in territorial marking, but the secretion from the inguinal gland is used only in odour marking of group members and young so it is not too surprising that it is of little use in territorial behaviour (Mykytowycz *et al.*, 1976). The boost to confidence provided by the background odorous environment may be quite common among those species which distribute their marks throughout the hinterlands of their territories. Sprainting sites in otters, while being attractive to vagrants and neighbouring residents alike, tend to cause a marked change in behaviour of the perceiver, resulting in its avoidance of the territory holder. If the confidence of an intruder overcomes its reticence to enter hostile land, it is usually the visual, rather than olfactory, threats of the resident which finally cause it to flee.

For those species maintaining a clearly defined scent-marked territorial boundary, frequent patrolling of it may serve to reduce aggressive tension in the border zone. Under conditions of fixed territorial occupancy with holders remaining tenants for quite long periods of time, neighbours will constantly be aware of the identity of each other. Territorial male lemurs, *Lemur catta*, respond by brachial gland marking much more to the odour of strangers than to the odour of territorial neighbours. Their response to neighbours' odours is typical of habituation and can be revived by the odour of a stranger. In the real world of territorial stability, a strange odour can mean only one thing – a threat to that stability which calls for immediate action (Mertl, 1975, 1976, 1977). Under experimental conditions of odour deprivation experienced by a territorial male lemur who can see but not smell his neighbour, there is a very marked increase in the conflict behaviours of approach and threat. This is marked by the deprived individual's ears standing out from the side of the head instead of lying flat. Such an animal is clearly highly stressed. The presence of odours, both familiar and unfamiliar, makes an individual more sensitive to visual cues than when it is deprived of odour. It is apparent that in these highly coloured and visual mammals, the integration of sight and olfactory signals is remarkably developed.

To the territorial male the possession of a territory is a double-edged sword with very obvious advantages. The maintenance and continued tenancy of a territory depends upon the holder's ability to exert his influence in every corner, and a very important function of scent marks is to make him feel he belongs in every quarter. The enhancement of confidence may be a more important widespread function of scent-marking than has hitherto been thought. As population density rises, the pressure on territories increases, and under these conditions it is necessary for the boundary marks to be constantly reinforced. This will reduce to an absolute minimum the chances of an intruder entering an occupied territory unknowingly, an event which would result in a chase or even a fight. Boundary marks seldom provide olfactory information only; the messages they convey, which relate to the dispersion of environmental resources, are of sufficient selective advantage to have been acted upon by selective processes influencing their additional visual and acoustic component.

6.4 Correlation between aggression-motivated behaviour and scent deposition

It will be evident from much of the above that odours are deposited in the contexts of behaviours which are themselves strongly influenced by aggression, and so far the discussion has covered the influence of odour in the expression of aggression itself, the structure of the social hierarchy, and the tenure of territory. Odours are often deposited on conspecifics for a variety of reasons, all of which have an aggressive basis and all of which are intimately intertwined with the social structure of the population. Odour marking of a partner of the opposite sex often occurs during courtship and, if the species is one with a pronounced and long-lasting pair bond, the outcome of such an activity has an important influence on population social stability. If the pair bond is short, the temporal influence of odour marking may be restricted. In its most dramatic form the marking behaviour of one sex by another is best seen in the rabbit; guinea pig, *Cavia porcellus*; mara, *Dolichotis patagon-a*; agouti, *Dasyprocta aguti*; chinchilla, *C. chinchilla*; and the porcupines, *Erethizon dorsatum* and *Hystrix leucura*; and it may occur in other species. As its excitement mounts, the male either runs round the female, sometimes jumping in the air or rearing up on his hind legs in front of her, and in so doing directs a jet of urine at her. It is not reported that this has any effect on the female one way or another and its function may simply be the release of the male's frustration (Ewer, 1968). Whatever the reason, the scent mark so applied functions as a label stating that the female is 'owned'; attentions by other males are not to be encouraged and anyway would result in wasted reproductive effort. If this is the true interpretation of sex partner marking, it can be seen that the aggressively motivated behaviour has a strong adaptive significance at the population level. Just as in territory boundary marking, the signal does not say 'keep off – this is my property'; rather it seeks to educate would-be fellow travellers.

An indirect means of urine marking a mate is seen in many of the members of the ungulate family Capridae. The process of self-urination by males has been mentioned previously in the context of the expression of reproductive vigour. This is likely to be its prime function, but during mating urine is incidentally transferred from the ventral parts of the male to the dorsal parts of the female, thereby signalling to other males that their activities may be more fruitfully redirected (Coblentz, 1976).

The passive marking of sex partners is probably quite common and probably occurs in all species with strategically positioned scent glands and an active courtship. Since all courtship involves a subtle interplay of sex drive-motivated approach and fear-motivated fleeing behaviour, the scent spread either actively or passively rapidly becomes a common scent and may act in part to help resolve this conflict in both partners. Once again it appears that the ability odour has to spread familiarity, and to boost self-confidence, is very great.

It is not just sex partners that are the target for scent deposition, for frequently offspring are treated in this way. Parent rabbits of both sexes mark their kittens with chin and inguinal gland secretions and the importance of this to the future survival of the young is great. Female rabbits are openly hostile to young that are not their own. They are able instantly to determine whether any particular kitten comes from their own or another colony and, while they merely harass young from their own colony, they hotly pursue and kill young from other colonies. The fact that it is the odour of the young which determines the response of the adults and not any features of their visual or acoustic output is seen when the kittens of a female smeared with odour from other rabbits are attacked and killed. Chin gland and inguinal gland secretions, but not anal gland secretions, trigger this response (Mykytowycz and Dudzinski, 1972). In the sugar glider,

Petaurus breviceps, a species of small arboreal gregarious marsupial, social harmony within the community is maintained by a single dominant male who anoints all the members of his colony, adults and juveniles alike, with secretions of a sebaceous gland complex overlying the frontal region of the head. The secretions of this gland complex are spread widely throughout the colony by a 'hugging' and rubbing behaviour and the common odour appears to ensure social well-being. If a colony member is removed and later repatriated, having been kept in isolation, it will be avidly inspected and energetically marked. If, however, it has been daubed with the secretion from a dominant male of another colony, it will be set upon and probably killed when repatriated (Schultze-Westrum, 1965). The same series of events has been observed in rats (Barnett, 1963), ground squirrels (Steiner, 1973), mongooses (Rasa, 1973), tree shrews (Martin, 1968), lemurs (Petter, 1965) and meerkats (Ewer, 1968), and probably occurs in many other gregarious species.

6.5 Summary and conclusions

Particularly in the mammals, odour deposition and olfactory perception play a substantial role in the social integration of species. Some odours appear to be associated with dominance, but it is not clear what role the better environmental conditions enjoyed by dominants plays in altering specific or general body odours. There is evidence from laboratory studies with mice that the odour of dominance is repellant to subordinates, but this is not universally true. In marmosets, for example, the odour of dominants is more attractive than that of subordinates. These differences, sharply revealed by conditions of captivity, may reflect the normal social grouping system utilized by the species. Mammals very obviously distribute scent around their territories, but such odours do not prevent intruders from entering occupied territories. The marks enhance the familiarity of the area to the resident, boosting his confidence and increasing his chances of victory in a conflict situation with a stranger. Scent-marking frequency is under the control of sex hormones and is linked closely to aggression-motivated behaviours. Because of this, it frequently accompanies courtship. In very gregarious species which live in a tight social group, group odours, often originating from a single male, are shared amongst group members. Like their role in territoriality, such odours serve to increase familiarity and reduce tension within the group. Juveniles are marked by one or both parents soon after birth and, as long as they retain the group odour, live in peace.

7

Alarm and defence

Animals need to defend themselves against physical attacks launched by predators and by others of their own kind. In nearly all species the ability to defend themselves is well developed and takes many forms; some are mainly structural, many are mainly behavioural. It is probably true to say that the ability is developed to its optimum under the conditions in which the species has evolved and may be unsuitable if the community structure changes dramatically. For example, Australian rats of the genus *Rattus* are noticeably more docile than European species of the same genus; the difference is because levels of predation in Australia are much lower than in Europe and the necessity for self-defence arises less often. To be forewarned is to be forearmed; part of the self-defence ability is the capacity to recognize an enemy while there is still time to take the appropriate evasive or defensive action. Since the selective pressures on self-defence tactics are likely to be among the highest experienced by the species, it is likely that all sensory channels are finely tuned to enemy detection but, depending upon the specific habitat in which the species lives, one sense may dominate. In day-flying birds, for example, visual signals give warning of danger; in night-flying birds the cues are probably acoustic. Very often detection of the presence of a predator will be associated with a particular behaviour which serves both to warn conspecifics of the danger and to indicate to the predator that its presence no longer can go undetected and that its chances of making a surprise attack are considerably diminished. For example, the mara, or Patagonian hare, *Dolichotis patagon-a*, repeatedly exposes its white lateral flashes when it is disturbed by a predator (Smyth, 1970) and the springbok, *Antidorcas marsupialis*, adopts its curious 'pronking' attitude during which it leaps into the air with its long white dorsal crest erected (Millais, 1895) (Fig. 7.1c, d, Fig. 7.2). The

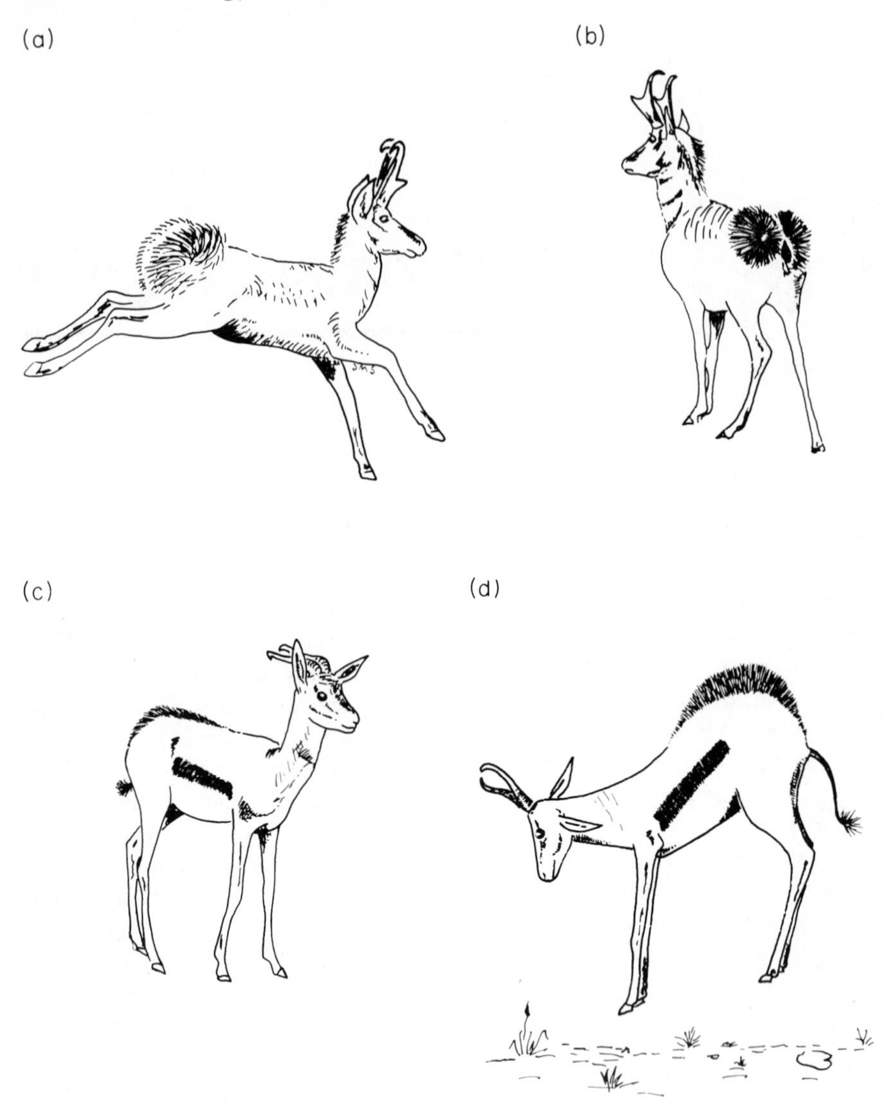

Figure 7.1 Combined olfactory and visual warning display in pronghorn, *Antilocapra americana* (a) and (b), and springbok, *Antidorcas marsupialis*, (c) and (d). (b) after Seton, 1927; (c) and (d) after Millais, 1895.)

distinctive wing dragging and crying behaviour of the lapwing, *Vanella vanella*, serves not only to warn the young to remain quiet and motionless but to draw the predator away from the locality of the nest. Accepting, fully, the integrated nature of the sensory detection of predators and that examination of the role of

Figure 7.2 Young springbok, *Antidorcas marsupialis*, from the rear, showing the width of the dorsal crest. (Photograph: D. R. Liversidge.)

one sense to the exclusion of the others is to some extent artificial, there are nevertheless a great number of vertebrate species in which the olfactory sense is clearly dominant over vision and hearing. Consideration of the role of odours in these species reveals some general trends which are potentially of the greatest significance to pest control and animal husbandry studies. This chapter will examine the role of odorous cues in (1) the detection of predators; (2) the transmission of alarm or warning signals to conspecifics; (3) active self-defence; and (4) the inhibition of attack by conspecifics.

7.1 Detection of the predator

Animals use all the means at their disposal to enable the detection of predators to be quick and accurate. Professor K. Lorenz was able to show that goslings reared in an incubator, and hand-tamed, respond dramatically to a cardboard silhouette of a hawk in flight even though they have had no experience of live hawks. The same goslings show no concern when the same silhouette is pulled along an overhead wire with the tail leading and now resembling a swan. This indicates that their ability to recognize the shape of a predator is inherited. Animal responses to the odours of their predators are not so clear-cut, and suggest that the precise course of the evolution of predator and prey may be important. One thing is clear, however. Although within any one predator species there is a large range of related odours associated with age, sex, dominance, etc., and demonstrably different to the members of the species, as far as the prey species is concerned there is an unmistakable species characteristic which cuts across individual differences. The mouse does not mind whether the hunting cat is a female in anoestrus or a dominant male; both represent a threat. Several laboratory studies have shown that the odour of a predator disrupts normal prey behaviour. The presence of a cat in or near a colony cage of rats brings about a 'freezing' behaviour which may last half an hour or more. If the cat is put in a large glass jar, however, so that olfactory contact is precluded, the rats go about their business as if no predator is apparent. Similarly, if the rats' nostrils are blocked with cotton wool, or if their olfactory bulbs are removed surgically, they remain apparently oblivious to the cat (Griffith, 1920). Such a clearly shown olfactory response does not tell us that the rats are responding any more to the smell 'cat' than to the smell 'novel odour', for all vertebrates show some sort of neophobia to new environmental components. In an experiment designed to test this point, thirsty rats were trained to run along a seven-feet-long trough to a reward box where water was available. If a cat was allowed to walk along the trough towards the water supply, unseen and unheard by the rats, the effect on the rats was as before – it took them five times as long to complete the journey to the water. But when a commercially available aerosol deodorant was used liberally all over the trough, no increase in running time was observed, indicating that the rats' response was to the specific nature of the cat odour and not just to any unexpected odour (Courtenay *et al.*, 1968; Mollenauer *et al.*, 1974).

Cats and rats have had a long evolutionary history in parallel – both have Old World origins. Like Lorenz's goslings it is possible to imagine that rats are genetically predisposed to cat odour. What is the outcome of the effect of the odour of a predator which has had no shared evolutionary history with a prey species? No response by the prey to the predator's odour should be expected. This turns out to be correct, although there are some interpretational problems. To the human nose the urine of the jaguar, *Panthera onca*, has a powerful, pungent, nauseous smell. The liberal daubing of this urine around the entrances to rodent live-catching traps set in an English meadow for short-tailed voles, *Microtus agrestis*, reveals that rodents are undeterred by the smell. Although the number of voles caught is the same as in a control trapping programme using clean traps, the total number of recaptures is somewhat reduced. This might indicate that the odour is slightly unpleasant, but it does not suggest that the urine contains an instantly recognizable component common to all predators and eliciting withdrawal behaviour in the prey. It apparently contains little information (author's unpublished observations).

So far the model is quite simple but, alas, there are further complications which upset the simplicity. Further field studies in the same English meadow, but involving the wood mouse, *Apodemus sylvaticus*, as well as *Microtus agrestis*, reveal that the response shown by one small rodent species to the odour of one of its main predators is not shared by others in the same community, yet both are subject to the predators' attentions. In this case the predator is the weasel, *Mustela nivalis*, and the effect of its anal gland secretion on the *trappability* of its prey is shown in Table 7.1. Although the diet of the weasel is mainly small rodents and young rabbits, the proportion of each type of food item in the diet varies from one place to another, no doubt in relation to the relative abundance of the prey species. Woodmice are reported as constituting 8.5 % (Day, 1968) and 50% (Walker, 1972) of the weasel's prey in two separate studies in Britain and, in the areas immediately surrounding the grassland meadow where the study was conducted, their abundance is quite high. In grassland habitats the woodmouse is usually non-resident. These results are puzzling not only because of the clear difference between the species, but also because the vole apparently is able to detect its enemy and show withdrawal behaviour. How has this evolved? What role does learning play? Does it not lead to imbalance in the finely adjusted relationship between predator and its prey? The long-term stability of the

Table 7.1 The effect of weasel odour on the number of rodents trapped in an old field, grassland habitat in southern England. (From Stoddart, 1976b.)

A Total rodent captures

	Total number of captures (including recaptures)		*Number in traps (tainted with weasel musk)*		*Number in clean traps*
	66		25		41
		$\chi^2 = 3.8788$		$p = < 0.05; < 0.02$	

B Differential effect of weasel odour on numbers of the woodmouse, *Apodemus sylvaticus*, and the short-tailed vole, *Microtus agrestis*, trapped

	Total number of captures (including recaptures)		*Number in traps (tainted with weasel musk)*		*Number in clean traps*
A. sylvaticus	32		14		18
		$\chi^2 = 0.50$		$p = > 0.05$	
M. agrestis	34		11		23
		$\chi^2 = 4.2353$		$p = < 0.05; > 0.02$	

interaction suggests that the prey must not become too good at escaping from the predator; neither must the predator become too good at detecting its prey. Clearly we know little of the functional behaviour of the system.

Californian ground squirrels, *Spermophilus beecheyi*, exhibit intense attention and defensive behaviour when they see a rattlesnake, *Crotalus viridus*, or a gopher snake, *Pituophis melanoleucus*, but perception of the odour of the snakes puts the behaviour into top gear. Since many snake–squirrel encounters occur in the rodent's burrow system, the olfactorily induced response may dominate the visual one. Interestingly enough, the response occurs in naïve squirrels and is not dependent upon encounter and learning (Henessy and Owings, 1978).

Reptiles depend heavily upon olfaction for many aspects of their ecology and the same is true for the detection of their predators. Even highly predatory species of snakes, such as rattlesnakes, have enemies, and the chief of these is the king snake, *Lampropeltis getulus*. The normal striking or defence posture for rattlesnakes to adopt is for the head to be raised and the anterior third of the body to be bowed out in a smooth loop. This occurs to the accompaniment of a rattling of the tail bones. If a rattlesnake is placed in an arena which previously held a king snake, or any other ophiophagous snake for that matter, a most unusual posture is assumed. The head is held pressed to the ground and the anterior third of the body is raised into an aerial loop which is thrown forward, lassoo-style, at objects in the arena. This peculiar defence posture is induced by odour alone – a rattlesnake deprived of the use of its Jacobson's organ, by the cutting off of its tongue, shows no defensive behaviour whatsoever when placed in a jar that is heavily scented with king snake odour. Rattlesnakes also react strongly to the odour of the spotted skunk, *Spilogale phenax*, but only if it is able to see the skunk. Commercially available skunk musk, thio-alochol *n*-butyl mercaptan, does not induce any defensive behaviour in rattlesnakes unless it is clearly associated with a mounted skunk skin. The rate of heart beat, however, increases markedly when the snake is presented with skunk odour, even though no whole body reaction is seen. When the rattlesnake's mouth, but not nose, is taped up so that the tongue cannot sample the air, the heart beat still increases but not quite at the same rate, indicating that both Jacobson's organ and the nose are involved in predator odour detection (Bogert, 1941; Cowles, 1938; Cowles and Phelan, 1958).

The site of production of kingsnake odour is the dorsal skin – ventral skin fails to elicit any reaction. Dorsal rubbings from king snakes of the eastern USA race are just as effective with western rattlesnakes as western race kingsnakes. This suggests a wide degree of odour equality or similarity in scent characteristics to which all potential prey respond even if the donor belongs to a different race that the prey would never, for geographical reasons, encounter. Rattlesnakes may have low powers of discrimination and be unable to distinguish between the odours of predators from which they have nothing to fear from those from which they have much to fear, but the alerting mechanism has a high sensitivity.

Fish, like snakes, appear to be capable only of a stereotyped response towards the odours of their predators and show no evidence of the capacity to ignore an odour which normally is associated with danger. Little work has been performed on regional differences in the diet of catholic feeders like pike, *Esox* spp., but it can be assumed that a substantial amount of local variation exists concomitant with prevailing ecological conditions. Yet the odour of pike, i.e., the water in which they have been held, exerts a powerful effect on many species of small fish which do not otherwise respond actively to the odour of other fish. The North American mosquito fish, *Gambusia partruelis*, for example, respond to most strange fish by orienting themselves in such a way that they avoid the area of attack, i.e., the area immediately above and in front of the stranger's head. They appear to do this visually. If the stranger is a pike, the mosquito fish respond by swimming upwards near to the surface of the water and adopting a stiff-finned, rigid posture. Their eyes darken and shaded bands appear underneath them. When in this entranced state they often show the behaviour from which they derive their common name; they jump into the air and skip along the water surface for sometimes considerable distances. The passing of a shadow over the surface of the water normally makes the fish dive rapidly, but when in this state they are little moved by such an event. This whole sequence is initiated if the water in which a pike has been held is introduced to a tank containing mosquito fish (George, 1960).

Pike prey heavily on minnows, *Phoxinus phoxinus*, and these small fish also respond to the odour of the pike. Depending upon the circumstances, minnows either flee and attempt to jump or spread their fins and sink slowly to the bottom where they remain inactive. Blinded minnows respond identically to sighted individuals, but fish rendered anosmic by surgery do not show any response to pike odour.

These sets of examples illustrate two fundamental concepts. The first is that, with the exception of the birds, recognition of a predator through the perception of odorous cues is a widespread phenomenon throughout the vertebrates, though it is best developed under ecological conditions which tend to preclude the dominance of other sensory systems. The second concept is that within the mammals there is evidence of some degree of modification of the stereotyped response, though it is unclear whether this is through learning or genetic inbreeding. Clearly more work is required in this area, but at the present it appears that the presence or absence of a particular prey response to the odour of a predator is a function of the feeding habits of the local predators, and this in its turn depends upon the distribution and abundance of prey species. A further intriguing question arises which bears heavily on the differences between visual and olfactory senses as perceptual systems. Young birds possess an innate fear of the silhouette of a raptor: do mammals possess a similar innate fear to a basic predator odour, if there is one, which may be lost subsequently if the ecological conditions so dictate?

7.2 Transmission of alarm or warning signals

When danger threatens one individual of a group, it is necessary for any alarm or warning signal produced to be transmitted throughout the group with the least possible delay. Odours are dependent upon air conditions for the rate at which they spread and would seem, under daylight conditions at least, to be potentially less effective than visual and acoustic signals. Alarm pheromones of insects are composed of smaller, more highly volatile molecules than those used in group identification or trail marking. Fading time of such pheromones is short and response thresholds are low – evolutionary correlates of their specific function (Wilson, 1971). There is an abundance of evidence implicating olfaction in the transmission of alarm within the vertebrates, but the involvement may only be secondary to an alarm posture, visual signal, or cry. Odours associated with fear or alarm may have taken on a communicative function from the changed physiological state associated with that condition. It is said that dogs, and other animals, can 'smell fear'; the production of aqueous sweat from the soles of the feet and palms of the hands during moments of extreme stress is a physiological phenomenon known to all and perhaps conveys the message. In some groups of vertebrates, notably the fish, physical damage to the skin of an individual is necessary for the release of the warning odour. In others, primarily the larger mammals, specialized sebaceous and sudoriferous glands secrete alarm substances which are released only when the animals are excited. Often these glands are associated with conspicuous tufts and crests of specialized hairs, often of a strongly contrasting colour, so that the release of the odour is functionally linked to a visual display. This is the most highly advanced form of olfactory warning seen in the animal kingdom, the most primitive being the necessary abrasion of the skin of one individual before its group mates respond.

During his studies on the European minnow, *Phoxinus phoxinus*, von Frisch had occasion to mark some individuals for later recognition by cutting the *nervus sympathicus* near the tail – an operation of a moment's duration which leaves a permanent black mark in the skin. Upon rehabilitation of the freshly marked fish to the stock tank, the unmarked fish reacted violently and formed a dense school ('Schreckreaktion'). He observed a similar effect when he released a fish that had become trapped underneath the metallic feeding tube. As it swam to join its companions, the group at first scattered rapidly then reformed in a tight mass. The precise sequence and progress of the fright reaction and its intensity depends upon a number of conditions including the strength of the initial stimulus. (von Frisch, 1938). As a result of this fortuitous observation, backed up later by many detailed studies, we have a clear understanding of the mechanisms of fright reaction and its evolutionary significance.

The site of production of the substance initiating fright behaviour is the club cells which lie within the epidermis and have no connection to the surface. Mostly they lie deep to the flask-shaped mucus cells whose secretions constantly pour out onto the surface of the skin. Such club cells are found only in the teleost fish belonging to the Gonorhynchiformes, and Ostariophysi and an alarm reaction has only been observed in these groups. Fright reaction substance-producing cells occur all over the body of these fish with the exception of sensory barbels. Although a complete chemical analysis of a wide range of alarm substances has yet to be completed, it appears that very few substances, and possibly only one – a pterin – is the active compound. Extract of minnow skin is highly effective in eliciting alarm reaction in the giant danio, *Danio malabaricus*, and in the characid, *Alestes longipinnis*.

The common occurrence of the fright reaction, and the occurrence of alarm substance-containing club cells in others which at the same time lack the alarm reaction (e.g., Salmonidae) has allowed an evolutionary analysis of the phenomenon to be made. Since the reaction is confined to the two groups Gonorhynchiformes and Ostariophysi, it is thought these have had a common ancestry. On osteological criteria the gonorhynchids undoubtedly arose from a salmonid ancestor, and it is thought that this latter group stood as a common ancestor for both. Present-day salmonids do not show a fright reaction but, alarm reaction substance-containing club cells have been found in fish belonging to the salmonid family Galaxiidae. The evolution of the fright reaction and alarm substance apparently occurred some time before the divergence of the Gonorhynchiformes and the Ostariophysi; in other words as early as the Jurassic or the Cretaceous. The Ostariophysi is undoubtedly a successful group of fish; there are 5000 to 6000 known species and the vast majority of fresh water fish belong to this group. It is tempting to think that this success, both in terms of the number of species and the length of time this group has been dominant, is partly due to the effectiveness of the warning signals.

An alarm reaction is not present in all members of the Ostariophysi, however, and is absent from the most aberrant forms like the blind cave fish, *Anoptichthys antrobius*; the highly predatory piranha, *Serrasalmus rhombeus* and *Rooseveltiella nattereri*; the spotted head stander, *Chilodus punctatus*; and the armoured catfishes of the family Loricariidae. Most ostariophysians are shoaling carp-like fish; those lacking the alarm reaction are the least carp-like in their life style (Pfeiffer 1974). It is also absent, on a seasonal basis, from several cyprinid (ostariophysian) species which show a nest-digging behaviour, or a behaviour in which a surface is cleaned prior to spawning. In these fish, such as the bullhead minnow, *Pimephales vigilax*; bluntnose minnow, *P. notatus*; fathead minnow, *P. promelas*; stoneroller, *Campostome anomalum*; common shiner, *Notropis cornutus*; and horyhead chub, *Nocomis biguttatus*, the males develop a ventral pad of thickened epidermis and connective tissue together with many mucus cells. As it develops under the influence of rising androgen levels in the blood, the alarm substance-containing cells disappear not only from the pad region but from the whole body. At the peak of breeding an extract of the skin does not elicit a fright reaction from conspecifics, but males with fully developed pads are still able to *perform* the fright reaction to alarm substances from, say, a breeding female or non-breeding male. The breeding season lasts from about May to August, after which time the pad gradually shrinks and the alarm substance cells reappear. The loss of production of alarm substance in these fishes is seen to be an adaptation to their abrasive spawning habits, preventing the release of alarm substance every time the nest is rubbed by the male (Smith, R. J. F., 1977).

An alarm-inducing substance acting in a parallel fashion has been noted in the amphibia; an extract from the crushed bodies of one or two tadpoles of *Bufo bufo* causes a panic flight in a shoal of test tadpoles when it is introduced into the shoal's tank. Bufonid tadpoles tend to form orderly schools and swim around in parallel rows. When extract is added (the active ingredient has been named bufotoxin), the orderly mass breaks up into wild confusion with most tadpoles swimming down into deeper water. Test tadpoles which have had their olfactory nerve transected by surgery do not respond to bufotoxin and behave normally even when their shoal mates are showing the alarm response. The sight of confusion occurring all around does not induce the alarm response in these anosmic tadpoles, a feature sharply contrasting with the response in minnows in which fish in one tank will show the behaviour if they can see fish in another tank which are being subjected to alarm substance. As larval fish do not show an alarm response to club cell secretion, it is unlikely that the response shown by larval amphibia has evolved from that of fish. The restriction of amphibian fright response to just one family – the only one showing pronounced shoaling of tadpoles – suggests it has evolved in complete isolation from its evolution in fishes. A wide range of anuran amphibians has been examined, but a fright reaction has been observed to occur only in the Bufonidae and is absent from all other families. It is worth noting that shoaling is primarily a feature of bufonid tadpoles; presence of and reaction to bufotoxin thus protects the whole shoal. As the tadpoles metamorphose, their reactivity weakens and their skins become less effective in eliciting the reaction in younger tadpoles (Pfeiffer, 1966; Hrbaček, 1950; Kulzer, 1954).

Snakes, lizards and other types of reptiles frequently release large quantities of musk from the postanal glands when they are disturbed. While it is likely this does, to some extent, fulfil a defensive role, musk has an effect on conspecifics much like the alarm substances described above. There is one case on record involving the king snake, *Lampropeltis getulus*, which is worth relating in detail as it makes a fundamental point (Brisbin, 1968). A young female king snake was held in the laboratory from the age of about two weeks and handled monthly for measurement and weighing for about 40 months. At this time an adult male king snake was brought into the laboratory and measured and weighed in the routine fashion. During this handling it emitted much musk which covered the work table. Although the table was thoroughly rinsed and wiped dry, it must have retained the odour of the musk, for when the female was laid on the table she reacted in an unprecedented fashion. Her tail tip was rapidly vibrated and within a half minute she herself emitted musk for the first time in her three years in captivity. It is possible that her response was to the sex or sexual condition of the musk donor, but her behaviour was typical of a snake in a highly disturbed situation. Although this is an isolated report of such an incident in the literature, the similarities with other alarm odour systems are quite striking.

Amongst the mammals, an olfactory alarm substance has been shown unequivocally to occur in mice and rats, but it is likely to be of more widespread occurrence. It appears to work in a manner similar to that in which alarm substances work in other vertebrates, although its site of production is not known. The fact that non-stressed rats and mice choose to avoid the alleyways in a maze along which a frightened or stressed rodent has passed up to eight hours before indicates that something is left behind, perhaps through the palmar glands, saliva, urine or faeces of the stressed rodent (Müller–Velten, 1966). Rats rendered anosmic through surgery show no aversion to the odour of stressed rats, while sham-operated subjects react just like intact rats (Rottman and Snowdon, 1972).

It appears not to matter whether stress is induced by applying electric shocks to the rodent, by its failure to win a fight with a conspecific, or by its confrontation with a predator. Any stressful situation appears to modify the odour envelope around the individual to such an extent that any previously important odorous effect of a particular individual is masked (Valenta and Rigby, 1968; Carr, Martorano and Krames, 1970). A problem besetting further analysis of this situation in laboratory rodents is that the precise moment of scent release, as well as the site of its production, are not known.

In many larger mammals the moment of release is not in doubt, because danger is signalled by the erection of superb crests of hairs, rump patches and other piliferous structures which are frequently brilliant white and in sharp contrast with the dark background hair colour. One of the most striking examples of the conjugation of visual and olfactory alarm signals is seen in the pronghorn antelope, *Antilocapra americana* (Fig. 7.1a and b). Observing the reactions of a buck to the unexpected presence of a dog, Seton (1927) states:

> It uttered no sound but gazed at the wolfish-looking intruder; and all the long white hairs of the rump-patch were raised with a jerk that made the patch flash in the sun. Each grazing Antelope saw the flash, repeated it instantly, and raised its head to gaze in the direction toward which the first was looking. At the same time, I noticed on the wind a peculiar musky smell – a smell that certainly came from the Antelope – and was no doubt an additional warning . . . in the middle of each disk is exposed a brown spot [the musk gland], from which a quantity of the musk odour is set free; and its message is read by those who have noses to read . . . even man can distinguish this danger-scent for 20 or 30 yards down the wind and there is every reason to believe that another Antelope can detect it a mile away.

The white hairs of the disc are about 10 cm long around the periphery and 5 cm long in the immediate vicinity of the musk gland, and are remarkably springy. When the disc is folded away, the entire length of the hairs lies bathed in the slightly yellowish secretion which perfuses the whole disc. Although the hairs are not specialized scent-broadcasting structures in any way, their erect posture allows for rapid dissemination on the wind.

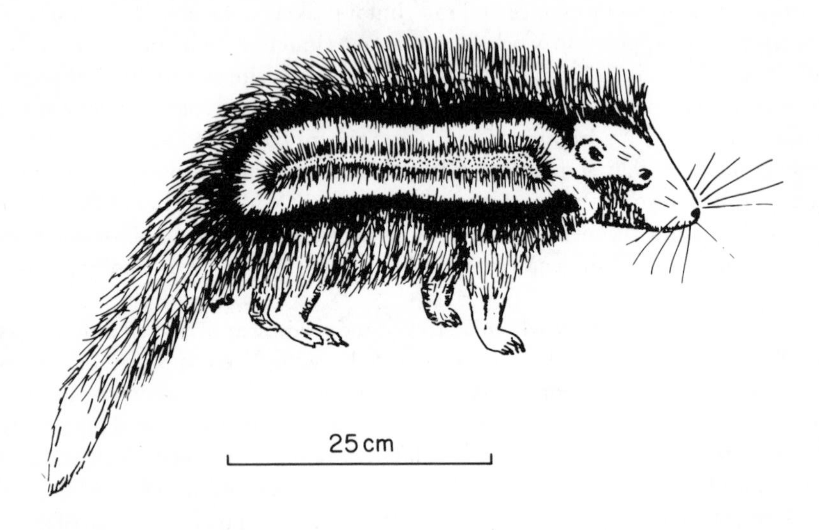

Figure 7.3 Crested or maned rat, *Lophiomys imhausi*. The erectile crest is fully raised exposing the lateral scent gland.

Another equally striking example is afforded by the crested or maned rat, *Lophiomys imhausi* (Fig. 7.3). This rather shaggy and unkempt East African rodent becomes suddenly transformed when alerted to danger, for the hair on each side of the body parts to reveal a black and white striped effect and a long, tapering scent gland. Arising from the course of this gland are a mass of short, stiff osmetrichia having a structure amongst the most complicated of any hair in the entire animal kingdom (Fig. 7.4a–c). Roughly spear-shaped, the medullary region of the centre of the hair is gouged into deep pits and the cortex is etched into a delicate filigree of cross struts. When the rat is alarmed, these hairs, filled with secretion like bath sponges, stand erect and can be bristled to provide an acceleration to the odour diffusion. The vastly increased surface area provided by the peripheral scaffolding of the hair further enhances dissemination. The raising and lowering of the gland's flanking hair serves as a powerful visual signal; the olfactory signal is probably no more than a back up, of particular use when the light conditions are poor.

In a somewhat analogous fashion to the adaptation seen in the pronghorn, the springbok, *Antidorcas marsupialis*, also utilizes a vivid white flash to signal

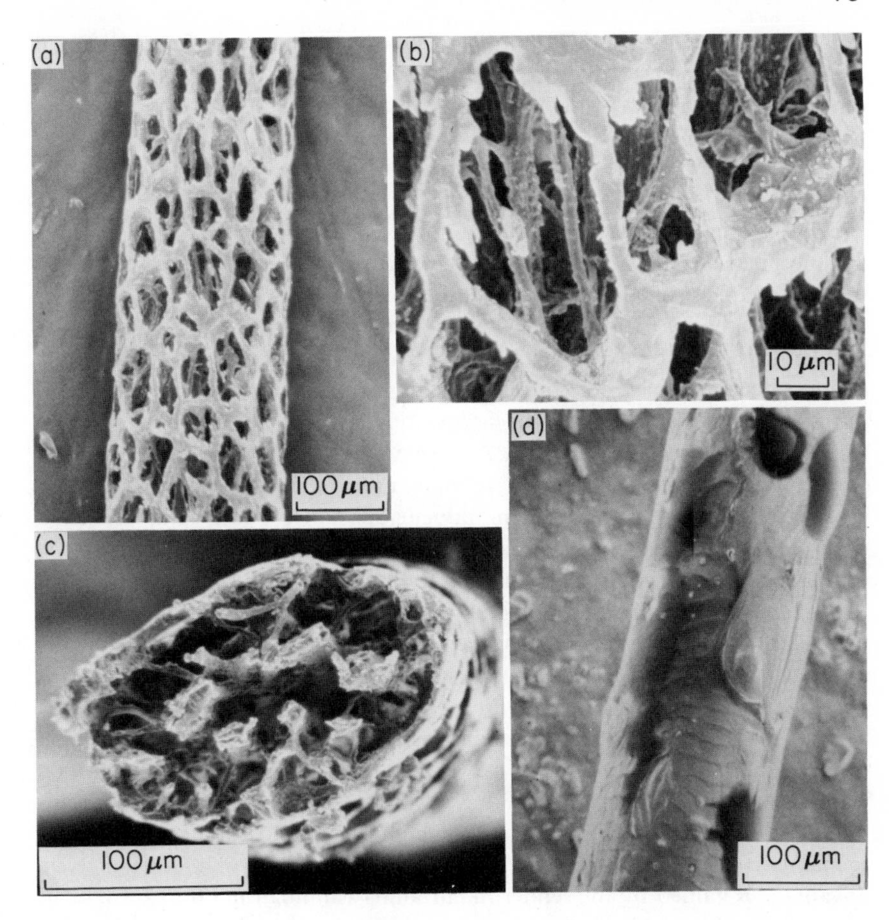

Figure 7.4 Osmetrichia associated with warning signals in two species of mammals.
(a) Scanning electron micrograph of surface of scent-releasing hair from the lateral scent gland of the crested rat, *Lophiomys imhausi*.
(b) The same; higher magnification.
(c) View of cross-section of the same.
(d) Scanning electron micrograph of dorsal crest hair of springbok, *Antidorcas marsupialis*. Note droplets of greasy secretion adhering to the hair surface.

warning. From the middle of the saddle to the base of the tail, and lying along the midline of the back, is a pouch (from which the antelope gets its specific scientific name) which is lined with pure white hairs measuring up to 16 cm long. When the antelope is alarmed, the crest is raised and lowered, sometimes very rapidly (Fig. 7.1c). To enhance the visual impact, the springbok indulge in 'pronking' behaviour, in which the back is arched during a high, straight-legged, leap into the air (Fig. 7.1d). The pouch is glandular, producing a copious watery secretion which covers all the hairs of the crest when the antelope is relaxed (Fig. 7.4d). As in the pronghorn, the hairs are not specialized as osmetrichia but their rapid movement effects speedy and effective odour evaporation. A complete analysis of crests, flashes and other distinctive pelage displays in mammals is long overdue; when it is completed it is likely it will show that most displays which are induced by fright not only inform the visual organs of conspecifics of danger but the olfactory organs as well.

Before taking the short evolutionary step from a consideration of the passive, or almost passive emission of an odour for the purposes of warning conspecifics to the purposeful use of odour as an active defence against predators, it is necessary to consider the evolution of olfactory warning signals, for they pose many problems. It is probably to the advantage of an individual to respond to the odour of damaged tissue of a conspecific as this can only signify danger, although in the case of the nest-digging species mentioned above a continuous response would clearly not be advantageous. As we have seen, in most ostariophysian and gonorhynchian fishes special cells are maintained having, apparently, the sole function of secreting a specific alarm-inducing substance, and the fright reaction occurs in response to the damage caused by a predator, not to the whole skin, *but to these special cells*. It can be assumed that the cost of producing and maintaining these cells is low because they are found to occur in blind cave fish and piranhas which lack any fright reaction to the secretions of conspecifics. What, then, is the selective advantage of these cells to the sender? If groups contain closely related individuals, such that the receivers are more closely related to the sender than the average degree of intrapopulation relationship, and the argument here applies quite strongly to social mammals as well as fish, the processes of kin selection could have fixed the occurrence of the cells within the population. The benefit-to-cost ratio is very high and therefore of a high selective advantage to the sender to maintain alarm substance cells. Again, assuming low cost of maintenance, an advantage is gained by the sender in curtailing cannibalism on its own, or its close relatives', fry. Certainly minnows become quite frightened after eating young of their own kind (Berwein, 1941). It has been suggested in a thoughtful discussion on the evolution of fish schooling that the primary function of alarm substance might be as a defence substance repelling predators (Williams, 1964). The vast amount of predation suffered by ostariophysians, and in particular cyprinids, rather argues against this hypothesis, but it must be remembered that in all ancient predator-prey systems coevolution has resulted in predators becoming specifically adapted to combat the defence mechanisms of their prey (Smith R. J. F., 1977).

Mammalian alarm signals differ from those of fish in that no tissue damage is required and there seems little doubt that they have evolved from the normal physiological reaction to stress related to the release of adrenalin. Because they are released prior to an attack, the sender may benefit by having all the members of his group react in a way that clearly informs the predator that he is under surveillance. In this way the chance of any individual being taken unawares is significantly reduced. The bright visual pelage display is likely to have a greater information content to the predator than the odour, unless the predator always hunts from downwind and there is always a wind blowing. Nocturnal prey species, however, may benefit more by an olfactory signal in response to a predator than by a visual display.

7.3 Active defence

Purposeful emission of scent for defence purposes has evolved rather infrequently in the vertebrates, although it is sometimes hard to distinguish between the production of an unpleasant-tasting or toxic secretion with one designed to warn off a predator before any contact occurs. Thus the cane toad, *Bufo marinus*, is characterized by enlarged parotid glands which are capable of ejecting their acrid secretion up to 20 centimetres. The smell of this is not as important as its irritating quality and the substance is particularly unpleasant if it should contact any mucus membrane. While this is a fine example of chemical defence, it is hardly an example of olfactory defence for the importance of the odour of the secretion is rather small.

Although most of the examples of odorous defence are found in the mammals, in which the means of defence develop early in life and remain active throughout the life of the individual, there is one known case of odour defence being used by a bird species, and only by a nestling at that. The bird is the hoopoe, *Upupa epops*, which is a native of southern Europe and a summer visitor to Britain. Like the hornbill, to which it is related, the hoopoe breeds in holes in trees or in ruined buildings. In sharp contradistinction to almost all other birds which maintain scrupulously clean nests, hoopoes make no attempt to remove faecal matter, and towards the time of fledging the nest smells appallingly of ammonia and rancid fat. The source of this odour is naturally produced metabolic waste which is removed by other birds so as not to advertise the presence of the nest to predators. Are hoopoes subject to higher predation on account of their unsanitary behaviour? The answer appears to be no, for the nestling hoopoe is endowed with the ability to produce, at will, an even more foul odour from the uropygidial, or preen, gland. Although carefully controlled field observations are lacking, it appears that this odour has a strong repellent effect on any inquisitive stranger. Young hoopoes make their first flight at about 30 days of age, by which time the uropygidial gland has shrunk from its maximum size, reached at about 12 days, to something approaching its adult size (Fig. 7.5a). The uropygidial secretion is not violently expelled; rather it flows to the glandular aperture where it is taken up by a ring of specially modified feathers which act like osmetrichia to dissipate the odour (Fig. 7.5b). This is not the only weapon in the young hoopoe's defence armoury; a predator penetrating the foul odour of the nest may have the liquid contents of the large intestine directed at it, accompanied by a hissing noise (Sutter, 1946).

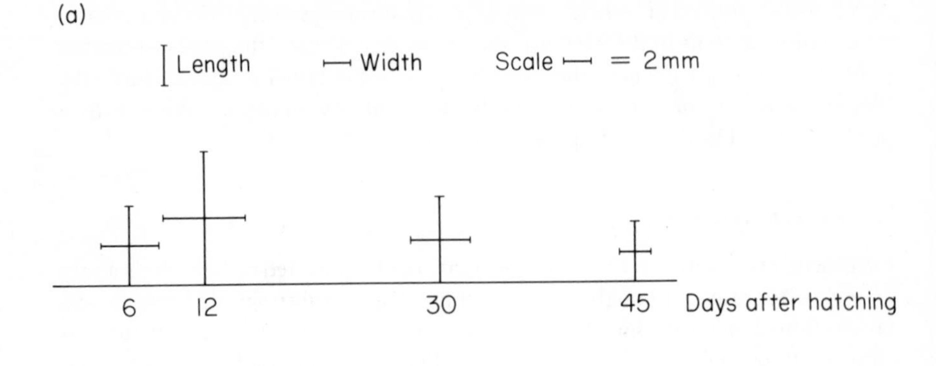

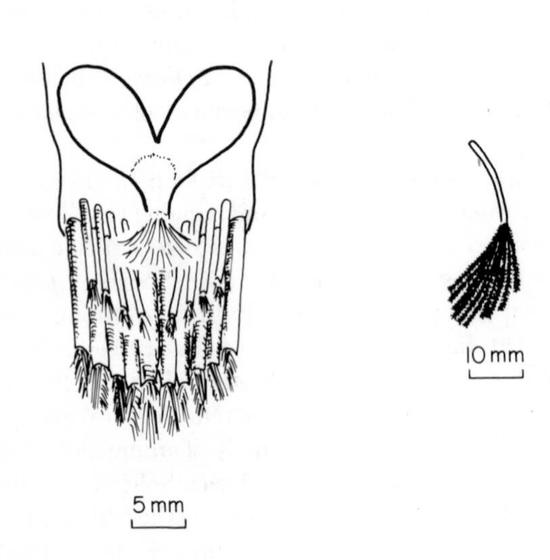

Figure 7.5 The uropygidial gland of the hoopoe *Upupa epops*.
(a) Length and width of one-half of the gland, against age. (From data in Sutter, 1946.)
(b) Position and size of the gland in a twelve-day-old nestling, and a specialized scenting feather from the glandular orifice. (Redrawn from Sutter, 1946.)

Before leaving the birds, it is interesting to note that birds belonging to the Procellariiformes may forcibly squirt stomach oil at an overinquisitive predator, but how this puts the predator off is unclear. It certainly has a foul, proteinaceous odour which lingers on clothes for a very long time, but it may merely serve to distract the predator's attention while the bird moves to safety, rather in the manner in which an autotomized lizard's tail fascinates the predator while the lizard effects its escape.

It is in the mammals that purposeful scent emission used for defence is most widespread, but even this term must be used with caution, for it is rather restricted to a few related families. Very many mammalian species have the ability to discharge urine, faeces and/or musk when suddenly led into a frightening situation, but this may be no more than the effect of the increased adrenalin level such situations beget. Only the unequivocal use of scent as an active repellent is considered here.

An interesting evolutionary progression in the modification of anal and perineal glands for defensive purposes is seen in the four small carnivore families, Mustelidae (weasels), Viverridae (civets), Herpestidae (mongooses) and Mephitidae (skunks). Although rapacious predators upon insects, birds, amphibia, reptiles and small mammals, they themselves fall prey to large cats, wolves and raptorial birds. The glands of all these families are under the control of striped muscle and can be emptied at will. Mustelids and viverrids void their perineal gland contents when threatened, but the scent is more deposited than ejected. The same secretion is used in territorial marking contexts and there appears to be no complex development of voiding behaviour. Mongooses again use their anal glands for territorial marking, but have developed some striking behaviours. The Indian mongoose, *Herpestes auropunctatus*, for example, backs up against a tree or other object and climbs backwards up it until it is standing on its forepaws. It does not invariably go through such a performance, however, and frequently squats to mark a twig or a stone (Gorman *et al.*, 1974). The marsh mongoose, *Atilax paludinosus*, spends much of its life up trees and has developed a striking behaviour for applying scent marks to branches. Marks are more often than not applied to branches above the mongoose's head which it reaches by performing a graceful 'hand-stand' posture (Ewer, 1968). This represents the ultimate in developmental complexity seen outside the family Mephitidae, and as the skunks are considered to be the youngest of the small carnivores, in evolutionary terms, it is thought their behaviours seen in defence scenting represent an extension of those seen in the weasels, civets and mongooses.

Skunks use their anal gland secretions very sparingly, and only as a last resort when all else fails to deter the attentions of a predator. The glands are reserved only for this purpose and are not used in territorial demarcation. Skunks are amongst the most contrastingly coloured of all mammals and this enables would-be predators to learn to recognize that they are obnoxious and distasteful in the same way that the yellow and black warning colour of bees and wasps acts to deter insectivorous birds. The mere adoption of the firing posture by a skunk is usually enough to halt a hopeful predator, but inexperienced and incautious young predators may trigger off a whole sequence. The first, and only, warning consists of the skunk jerking its rear end around to face its adversary and raising its tail clear of the perianal region. There are no further warnings. If this does not bring about the required effect, the tail is raised even higher and a stream of foul-smelling liquid is ejected from the anal glands outwards and upwards for a distance of about four metres with great accuracy. The striped skunk, *Mephitis mephitis*, arches its back prior to ejection so that the trajectory of the scent stream is matched to the height of the predator (Fig. 7.6a) while the spotted skunk, *Spilogale gracilis*, adopts a hand-stand posture astonishingly similar to that observed in the marsh mongoose (Fig. 7.6b). Because the hand-stand posture raises the perianal region high off the ground, it is possible it has evolved in relation to the size of the predators of the spotted skunk (Johnson, 1921). This is an interesting evolutionary point which appears not to have been taken up by ecologists and ethologists.

Figure 7.6 The firing posture adopted by two species of North American skunks
(a) *Mephitis mephitis*;
(b) *Spilogale gracilis*. (Redrawn from Bourlière, 1955.)

7.4 Protection from intraspecific attack

During its period of infancy, before it can lead a free and independent life, a young animal needs to be able to inhibit any aggressive action its parents, or any other adult, may show towards it. Or, to put it another way, it may require *not* to release any aggressive behaviour. A whole range of signals are used, from the 'infant-schema' of visual signals (Lorenz, 1943) to the ultrasonic cries made by infant rodents which may placate their fathers as well as acting as distress calls when they are cold or out of the nest (Sewell, 1967). It is hardly surprising to find that scents play a part in the provision of this protective shield. In the mammals two basic mechanisms are employed: the provision of scent by the mother who must apply it to her offspring, and the intrinsic production of scent by the offspring themselves. To take an example of the first type, female tree shrews, *Tupaia belangeri*, rub the product of their chin glands upon their youngsters and this affords them complete protection from attack by other adults. If the tree shrew population should rise too high, however, the mild stress this induces in the adult females causes the chin glands to atrophy. Lacking the protective scent, the youngsters quickly succumb to the attacks of all the neighbouring adults (von Holst, 1969). This seems an effective means of regulating the population size. The mouse affords a good example of the second type, although there is a suggestion that slightly different means of protection are applied to young of the two sexes. The urine of a juvenile female, like that of an adult or even ovariectomized female, has a marked depressive effect on male aggression, so the active ingredient cannot be associated with the ovary or any other part of the mature reproductive system (Dixon and Mackintosh, 1975). When the urine of a juvenile male is daubed onto one of a pair of adult males, there is no change in the level of aggressive interaction, while a similar daubing with juvenile female urine effected very significant reductions in aggressive interaction. But when juvenile males are placed in the territories of adult males they are attacked far less frequently than control adult male intruders. So while juvenile females are protected solely by intrinsic urinary compounds, juvenile males are protected by a complex of factors, not primarily olfactory (Dixon and Mackintosh, 1976).

As young male rodents develop into adolescents and then leave the nest to become independent adults, the quality of their scent gland secretions undergo dramatic changes. This is most clearly shown for the European yellow-necked mouse, *Apodemus flavicollis* (Fig. 7.7) (Stoddart, 1973), but also occurs in *A. sylvaticus*, *A. microps*, *A. gurkha* and *Gerbillus campestris* (Stoddart, 1977). Mice of the genus *Apodemus* have a distinct sebaceous gland lying along the ventral aspect of the proximal third of the tail, which contains a great many substances in the adult male. Juvenile males, i.e., males prior to sexual maturity, show a highly reduced spectrum which much more closely resembles the secretion of their mothers than their fathers. Adolescence, when the testes have commenced their descent but are not yet fully scrotal, lasts about three days, and secretions taken at this time show marked changes. Many of the substances which are quantitatively important in adult males, but lacking in adult females and juvenile males, start to appear in the spectrum. Although every step in this transition has not been recorded, it appears that harmony within the nest occurs when the juvenile males produce secretion which resembles that of their mothers. Once the secretion takes on male characteristics, at the onset of sexual maturity, the young males are forced to leave the nest. This idea of the juvenile males hiding behind their mothers' olfactory apron strings may be found to extend to a

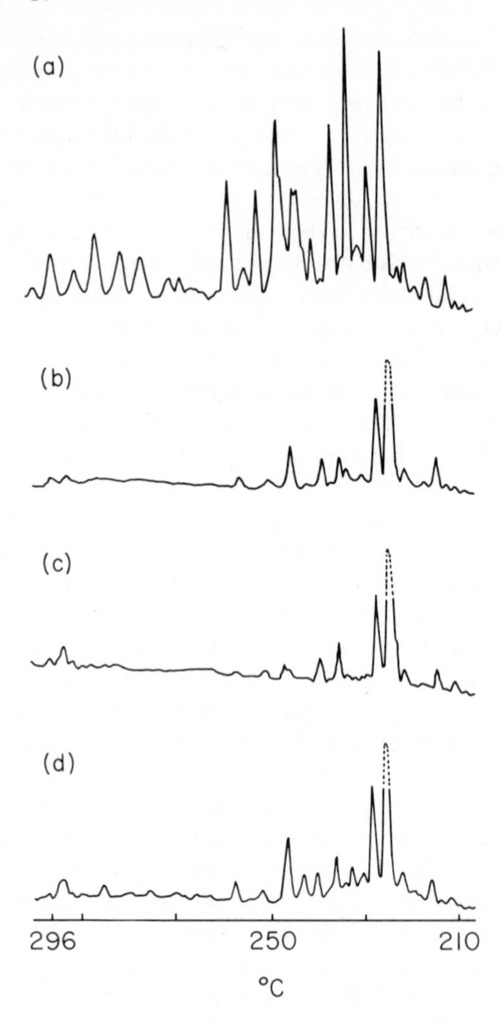

Figure 7.7 Gas chromatograms of caudal organ secretions from the yellow-necked mouse, *Apodemus flavicollis*. (a) adult male; (b) adult female; (c) juvenile male; (d) adolescent male. (Chromatographic conditions as in Figure 5.1.) (Modified from Stoddart, 1973.)

wide range of mammals, particularly those small species with a short weaning time where rapid production of litters is the rule.

Many ungulates are born almost scentless, particularly those altricial forms which do not need to run with the herd within a few hours. This is thought to protect them from the food-seeking attentions of predators while they rest in bushy thickets by day (Kiley-Worthington, 1965). Although no chemical or other analytical studies have been made, the idea seems attractive. It is possible, however, that the absence of odour reduces any tentativeness of approach by the mother. Such a phenomenon is seen in species like the waterbuck, *Kobus defassa* (which is noted, incidentally, for its oily, musky-smelling skin – the French common name of this species is '*Cobe onctueux*'). Lacking the constant visual and acoustic contact with the young that occurs in migratory, precocial species, the absence of any body odour emerging from the young may help facilitate the parent-offspring reunion by reducing the mothers' natural apprehension towards conspecific odours.

7.5 Summary

Perception of odours arising from predators elicits a wide variety of response behaviours in vertebrates. Rodents show an aversion to places frequented by their predators, and the stress so caused results in they themselves producing an odour which further influences the exploratory and locomotory behaviour of others. Anosmic individuals show no change in behaviour. Snakes respond to the odour emanating from the dorsal surface of their ophidian predators and themselves produce copious musk from their cloacal glands. This might act as an intraspecific warning odour as well as a predator repellent. In aquatic habitats the odours of certain predatory fish cause their prey to adopt escape tactics. If fish belonging to the Ostariophysi and Gonorhynchidae are injured by a predator, a special, though not species-specific, alarm-inducing substance is released from the ruptured dermal club cells. A substance having a similar effect is released when the skin of bufonid tadpoles is abraded. In both fish and tadpoles, artificially induced anosmia results in a lack of response, and in snakes surgical removal of the tongue dispenses with the fright behaviour. The contrastingly coloured tufts of hair on large mammals which are used for visual communication of danger are associated with glandular skin patches. In some species, e.g., the African crested rat, the hairs are modified into wicks for the broadcasting of the secretions.

The purposeful use of scent as an anti-predator repellent is seen only in the mongooses and skunks, in the latter group of which it reaches its highest level of functional development. In rodents either the presence of aggression-inhibiting substances or the absence of aggression-promoting substances appears to afford juveniles protection from attack from parents and other adult conspecifics. It is not known if this is generally applicable to the young of vertebrates of other classes.

8

Olfactory navigation and orientation

In preceding chapters the role of olfaction in influencing a behavioural response or an ecological relationship has been clear-cut, but in this chapter, which deals with odour cues in navigation and orientation behaviour, there is rather little evidence of an unequivocal nature. This is not because olfaction plays only a minor part in guidance behaviour, but because vertebrate animals utilize all their senses for this important task. If an experimentally blinded animal finds its way back home, the experimenter has not demonstrated that it normally does not use visual cues; rather he has demonstrated that it is able to gather enough information without visual cues. The selective advantage of a multichannel control system seems obvious since the environment through which a migrating animal has to pass is subject to change in both predictable and unpredictable ways. The change from day to night may necessitate in the sense most relied upon shifting from the eyes to, say, the nose, but so may the occurence of an unexpected fog bank. Working. in full integration with one another, the capabilities of each sensory system may be enhanced significantly and this may explain why the experimentally demonstrable sensory capabilities of migrating species sometimes seem inadequate for the task. In the light of the persistent mystery of animal navigation, it is surprising that so little attention has been paid to the part played by olfaction, but perhaps this is a legacy of the anthropocentric nature of the early work on behavioural ecology.

The object of this chapter is to highlight the known involvement of olfaction in both the long-distance navigational ability and in the short-distance homing orientation of vertebrates in order to achieve two things. First, it is hoped to indicate that an olfactory navigation system of quite incredible sensitivity occurs in one class of vertebrates and most probably exists in another, for which there is much circumstantial evidence though no proof. Second, it is planned to discuss the problems of interpretation of the results of experiments designed to illustrate the role of scent in homing behaviour in an endeavour to stimulate more critical research. Some speculation is necessary in the examination of this topic, but it is considered justifiable when it is directed toward the solution of long-standing puzzles and is not too outrageous in its nature.

Ecological and ethological research is often necessarily of long duration, and critical experiments of the kind beloved by physiologists and biochemists on the navigational ability of vertebrates are either extremely difficult or impossible to conduct. We should not expect rapid or unequivocal results in this field, although the use of implanted radio beacons and elegant tracking devices is already reducing to more manageable proportions some of the greatest logistical problems. In the meantime we must be content with some degree of conjecture in the analysis of the phenomenon of navigation and homing in vertebrates.

8.1 Olfactory navigation

Fish live in an environment in which there is potentially a much higher number of perceptible substances available than the number a terrestrial species might expect to encounter. This is because terrestrial creatures are able only to perceive substances which are volatile, and this depends not solely upon the molecular weight and vapour pressure of the substance, but also upon climatic conditions, but to fish, any even partially soluble compound is perceptible, irrespective of its molecular weight. In spite of this natural advantage, many fish have olfactory systems of almost unbelievable sensitivity. The eel, for example, can detect concentrations of β-phenylethyl alcohol as low as 3×10^{-18} M (Teichmann, 1957). This concentration can be achieved by diluting 1 ml of the alcohol in a volume of water equal to 58 times that of Lake Constance, and this means that no more than three molecules of the substance will be present in the olfactory sac of the eel at any one time. Yet this is enough for a response to be elicited. The odour hypothesis for home stream location by migrant salmon returning to their natal site for spawning demands sensitivities of similar orders of magnitude, and, although such powers have only been demonstrated experimentally for the eel, that is no reason to suppose they are lacking in other fish which undergo extensive migrations.

The basis for the odour hypothesis for home stream location has been painstakingly assembled by A. D. Hasler and his co-workers in Wisconsin who for many years have studied the breeding migrations of salmon (Hasler, 1966). Several species of salmonids exhibit this phenomenon, notably the Atlantic salmon (*Salmo salar*), Pacific salmon (*Oncorhynchus keta*, *O. kisutch*, *O. nerka*, *O. gorbuscha*, *O. tschawytscha*), and the rainbow or steelhead trout (*S. gairdnerii*). These anadromous species spawn in freshwater headstreams where, after hatching, the young fish remain for between a few weeks and two years, depending on the species, before heading off out to sea. They grow rapidly in mid-ocean feeding off the rich zooplankton at the edge of the continental shelf and, after two to seven years when they become sexually mature, start the long migration back to the parent stream for spawning. Adults of most species die after a single spawning but Atlantic salmon return to the ocean and may spawn several times over several years. Although some evidence has been assembled which suggests that olfaction plays a role in the oceanic phase of the migration (Bertmar and Toft, 1969), the experiments which have been performed have lacked adequate controls and it would be premature to conclude that olfaction is involved in this phase. Other cues, such as celestial configurations, polarized light patterns, and electric and magnetic field characteristics are all thought to play a part (Harden-Jones, 1968; Forward *et al.*, 1972; McCleave *et al.*, 1971; Rommel and McCleave, 1972), but consideration of the oceanic phase is excluded from this account because of the absence of much data of any kind. Long ago a distinguished English naturalist suggested that salmon responded to the 'smell' of their home stream and used it as a homing beacon (Buckland, 1880). Leaving aside for the moment the origin of the smell, the concept of odour homing can be broken into three main sections. Firstly, there must be a characteristic and very long-lasting odour associated with the stream. Secondly, the fish must be able accurately to discriminate between home stream and other stream odours, and thirdly, the characteristics of the home stream odour must be indelibly imprinted onto the brain of the fish so that it may be recalled after an interval of up to several years. In operant conditioning trials using *S. gairdnerii*, Hasler demonstrated that water taken from two small streams, each of which flowed through different soil and topography conditions, could be distinguished readily by trained fish. Furthermore, the distinction was made equally easily at all times of the year. The active ingredient was found to be organic in nature, for rehydrated ignited evaporation residues provoked no response. In a series of field trials conducted around a fork in a tributary system, mature coho salmon (*O. kisutch*) were trapped after they had entered one arm. A number of fish were tagged and about a half of them had their olfactory pits blocked with cotton wool. All fish were then transported three-quarters of a mile below the junction, and a short while later traps were operated on both rivulets above the junction. Although the sample sizes were small, it appeared that the majority of the untreated controls made the same choice at the fork as on the first occasion, but the treated subjects entered both forks almost at random. Such behavioural experiments are fraught with methodological and interpretational difficulties, but the olfactory navigation hypothesis gains support from studies of the electrical activity in the olfactory bulb of migrating fish. When the olfactory sacs of an anaesthetized salmon are irrigated with some home headstream water, a rapid burst of electrical activity is shown by electroencephalography. No such burst occurs when water from an unrelated stream is used (Fig. 8.1). Since all other conditions are identical, the only variable is the quality and quantity of

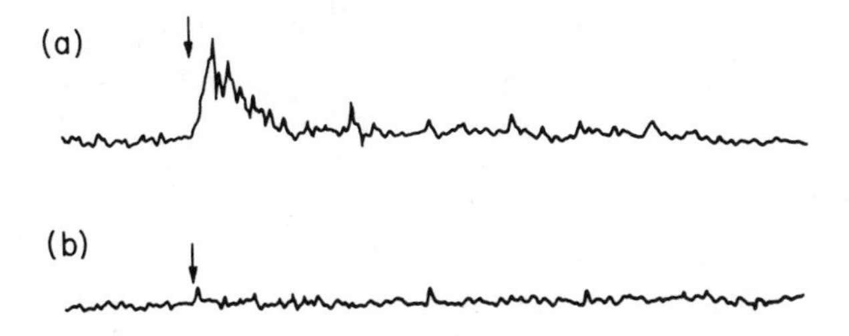

Figure 8.1 Electroencephalographic response of the olfactory bulb of the chinook salmon, *Onchorhynchus tschawytscha*, to home stream (a) and non-home stream (b) water. Arrows mark moment of perfusion of the olfactory sacs. (Redrawn from Ueda *et al.*, 1967.)

dissolved solutes in the two waters (Ueda *et al.*, 1967).

Quite an amount of effort has been expended to ascertain the nature and origin of the homestream odour. It has been shown to be organic, and until recently was thought to result only from the floristic and geological characteristics of the headstream catchment. This seems quite plausible, since it is quite within the known olfactory capability of fish to recognize, and respond to, waters in which certain plants have been immersed. Blunt-nose minnows, *Hyborhynchus notatus*, for example, can be trained to differentiate between water in which a sprig of a test aquatic plant has been rinsed and control water (Hasler, 1966). The olfactory system makes this discrimination, for cauterization of the olfactory lamellae with a hot wire eliminates this ability; but as with many such studies the lack of wholly adequate controls invites the exercise of some caution over the interpretation of the results. Furthermore, although a particular sensory system can be shown to be able to provide the individual with adequate information under laboratory conditions, this does not mean that the system is utilized in an identical fashion under natural conditions. Since a wide variety of watersheds with differing floristic characteristics support salmon runs, it is rather perplexing why some river systems are ignored altogether by salmon. Recently evidence has come to light which implicates other salmon as the source of attraction. The rivers Usk, Wye and Severn in the West of England have long been regarded as major salmon rivers. Into the southern edge of the Bristol Channel flows the river Parrett and a small commercial fishery has existed at the river mouth for some years (Fig. 8.2). The Parrett is not a salmon river, and in most years the total catch is about 100 fish. It is assumed that the fish taken are members of the Usk, Wye and Severn populations which are searching for their own headstreams. In the spring of 1953 some 25 000 salmon ova (from rivers in Scotland, Iceland and Wales) were introduced into a tributary of the Parrett, followed in the three successive years by a slightly lesser number. In the summer of 1953 the commercial fishery had a bountiful harvest and in 1956 it rose to about five times its level prior to the experiment (Fig. 8.3). Tagging experiments on the Usk, Wye

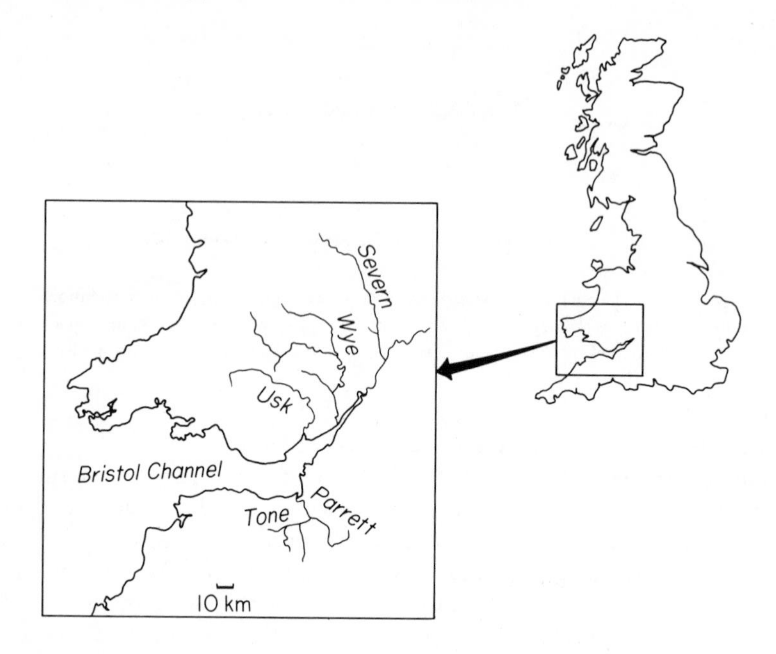

Figure 8.2 Map showing the location of the main salmon rivers in southwestern England, and the River Parrett and its tributaries.

and Severn rivers indicated that many of the fish taken on the Parrett originated not from the transplanted ova but from the other populations (Solomon, 1973). Similar evidence is reported from studies with char, *Salvelinus alpinus* (Nordeng, 1971). A number of artificially fertilized eggs of the char taken from the Salangen river in northern Norway were transported by air some 2000 km south to a hatchery at Voss. After hatching, the young fish were kept at Voss for four years before being flown north to the river system of their origin. Some 10 km distant from the estuary of the Salangen river is the mouth of the Løkesbotn river, though both empty into a sheltered inner bay. Of those fish released into the Løkesbotn and subsequently recaptured, 67 per cent were caught in the Salangen and 34 % in the river of their release. Of another batch released in the Salangen river, about 95 per cent of those recaptured were taken in the Salangen. As all the eggs came from the Salangen in the first place, these results are interpreted as demonstrating that the young fish respond, inherently, to the odour of the river which contains their relatives. That it is the presence of their *relatives*, and not some other character of the water that attracts the young fish, is indicated by an experiment performed with some of the Salangen stock maintained at Voss. Some 200 smolt were removed from the hatchery and released about 5 km downstream from the holding tanks which still contained many char. Within a few days, a large number of the young char were observed

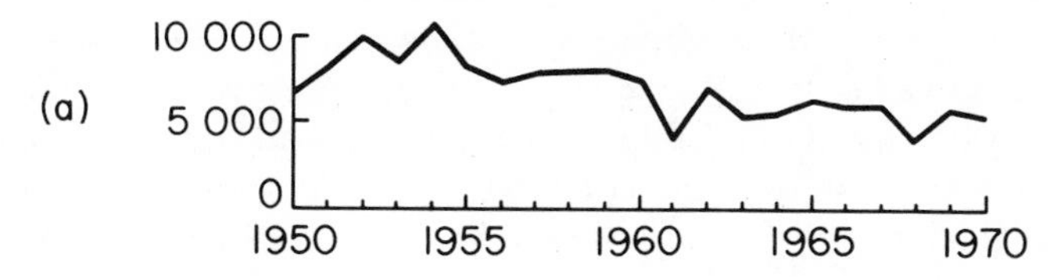

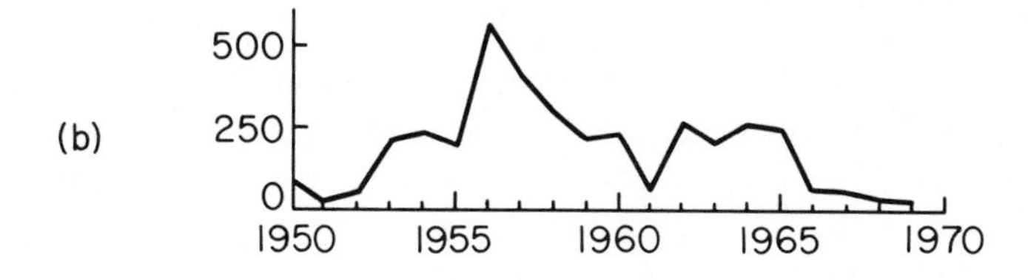

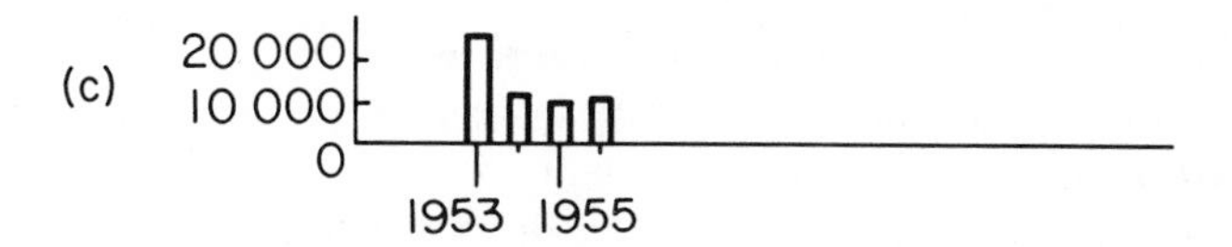

Figure 8.3 Diagram showing
(a) annual commercial catch of salmon for the Usk-Wye-Severn tidal fishery;
(b) annual commercial catch for the Parrett estuary fishery;
(c) numbers of salmon ova planted in the River Tone, a tributary of the Parrett.
(From Solomon, 1973.)

to gather at the outlet from those tanks. They apparently did not disperse throughout the Voss river system, although they were free to do so.

Whatever the source of odour, and there may be more than one source involved, there is clear evidence that the homestream odour is remembered. In the experiments with char outlined above, learning of the homestream odour was not possible since the ova were transported prior to hatching. In other experiments involving coho salmon (*O. kisutch*), some smolts were artificially exposed to morpholine, which neither attracts nor repels fish, at concentrations of about 5×10^{-5} mg/l (Madison *et al.*, 1973). Both exposed and unexposed control fish were subsequently stocked some distance from the hatchery. Eighteen months later, during the first adult spawning run, morpholine was dripped into a stream. In four separate experiments, 10 times the number of scent-exposed fish as controls were recaptured in the scented stream. In a fifth experiment, morpholine addition was abandoned and at once the numbers of exposed and unexposed fish entering the stream became approximately equal. It would appear that smolts undergo a sensitive period which occurs at the start of the downstream migration during which they become imprinted on the odours of their environment. If the smolting stage is allowed to pass before the young fish are released into a headstream, the subsequent recapture rate is far lower than for an identical batch of fish that set off downstream at the right developmental time. It is unlikely that two distinct mechanisms of odour recognition occur in two species of salmonids; the different experimental techniques employed results in the char smolts perceiving, and remembering, the smell of conspecifics while the coho smolts respond to the odour of their headstream. Presumably both mechanisms operate all the time and both exert a strong selection pressure on the mature fish to seek the company of others.

A physiological problem is raised by these experiments. Given a point source of an odour and the presence of currents of varying intensities, a fish could only home in on that point by tropo- and klinotaxis; i.e., it would have to perceive spatial differences in concentration between the two narial sacs, as well by the radiation characteristics emanating from that source. Tropotaxis might occur when the odour concentrations are high, but at levels of 5×10^{-5} mg/1 it would seem that the close mutual proximity of the nares precludes such perception. As has been discussed earlier (p. 26), in most fish there are a series of physical and behavioural adaptations which enhance the assymmetrical sampling abilities of the nares. Although these are no experimental data yet, a further look at the mode of action of the external nares may help resolve this paradox.

Every bit as fascinating as the migrations of fishes are those of marine turtles. The green turtle, *Chelonia mydas*, for example, annually commutes between the eastern coast of Brazil, where it feeds on the lush growth of turtle grass, and tiny Ascension Island, some 2250 km distant. After breeding on the sandy shores of the island, they return to Brazil followed, after an unaccountable delay of one year, by their young. During the non-breeding season, turtles inhabit the Brazilian inshore waters from as far north as Fortaleza to Rio de Janeiro in the south. It is widely accepted that the separation between feeding and breeding grounds arose as a result of continental drift, and the continued existence of the turtles says a lot for their powers of adaptation. To a marked degree their success has depended upon their continued ability, generation after generation for millions of years, to locate their breeding ground as it slowly drifted eastwards. Although there are few experimental data from which to seek an explanation for this ability, careful observation of what actually happens to turtle populations during their life cycle indicates that an olfactory hypothesis has to be taken seriously. Young hatching turtles, just like hatching salmonids, become imprinted upon the odorous characteristics of their natal surroundings. Two or more years later they set out from the feeding grounds on their first breeding migration and the source of those long-remembered odours serves as a homing beacon (Koch, Carr and Ehrenfeld, 1969). This hypothesis seems more sound than one based on celestial or solar navigation because of the ease with which initial imprinting would occur, and odour imprinting occurs widely throughout the vertebrates. For a visual system to work, an assumption would have to be made that the coordinates of the natal site are remembered at the time the young turtles leave the island, and that later, as adults, they are able to apply those cues in reverse. Furthermore, although Ascension Island lies only about 8°S and experiences little difference in the sun's position throughout the year, some of its breeding stock come from as far as 20°S; for a visual navigation hypothesis to work would necessitate the ability of adult turtles making due allowance for their final position north or south of Ascension's latitude at a different time of year from the one at which they first made landfall.

Returning to the odour hypothesis, although the target is small – Ascension Island is only 8 km wide – only a small amount of specific attractant substance is needed. This is because of a well-defined 50 m deep thermocline which keeps the pelagic waters quite distinct from the deeps. The prevailing pelagic currents flow from east to west, and in washing past Ascension effectively treat its own specific effluents as emanating from a point source (Fig. 8.4). Diffusion of an hypothetical odour beacon occurs in a plane, and the time available for diffusion in a north–south direction is directly proportional, assuming a constant current speed, to the distance covered by the current in an east–west direction. Application of the Fick diffusion principle, derived to apply to molecules under the action of Brownian movement and therefore not ideal for the specific circumstances associated with oceanic currents, indicates that the dilution of the island odour at the point when

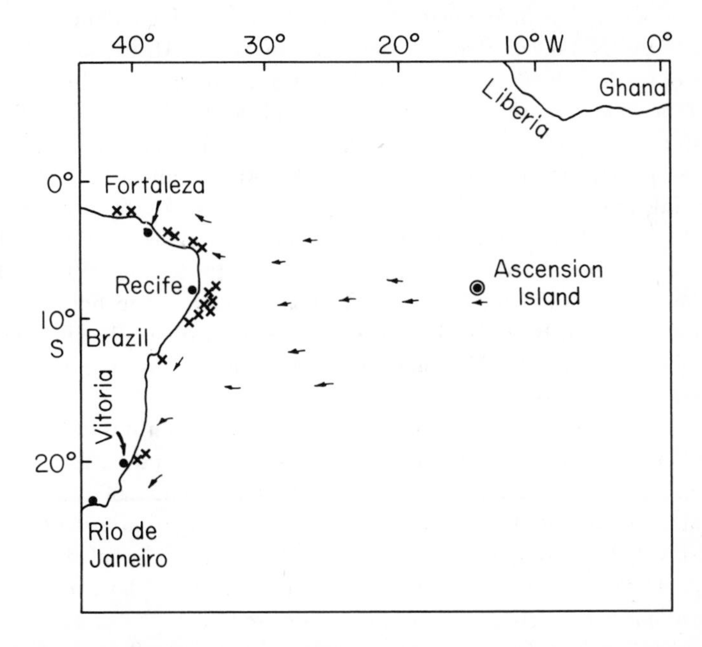

Figure 8.4 Map showing prevailing surface currents in the mid-South Atlantic, and points (x) at which green turtles, *Chelonia mydas*, marked on Ascension Island were recaptured. (Modified after Koch, Carr and Ehrenfeld, 1974.)

the current hits the mainland of Brazil is of the order of $\frac{1}{100}$ to $\frac{1}{1000}$. These projections at least indicate that olfactory orientation is possible, but the evidence is little more than circumstantial at present. Navigation from west to east would necessitate that the turtle is able to detect when it is no longer in an odour stream of the same concentration as previously. According to the theory, a turtle near the Brazilian coast would have to swim northward or southward for about 80 km in order to experience a tenfold increase in odour concentration – a task of some hours duration and one which necessarily requires an acute olfactory memory. Further, gradient comparison necessitates an avoidance of sensory fatigue, but this could occur by the turtle diving deeply to leave the thin upper layers of water containing the active substance – an activity which quite commonly occurs.

The argument thus far suggests that olfaction is important for the whole length of the journey, unlike in fishes in which only the final pinpointing of the home site is left to the nose. Turtles also appear to use their noses during the final stages of the journey, even after they have made landfall. As the females drag themselves out of the surf and up the wet beach they repeatedly press their snouts onto the sand for a few seconds. While such behaviour could be tactile (Carr, 1965), it appears to be an odour sensing. Female loggerhead turtles, *Caretta caretta*, normally walk up the beach with their noses buried, like ploughshares, in the sand, as if looking for some specific set of conditions. In Tongaland, an isolated portion of the Natal coastland in southeastern South Africa, large breeding colonies of turtles, chiefly loggerheads, have traditional nesting sites. Although much of the 25 km coastline is apparently suitable for turtles, the bulk of nesting occurs in two distinct areas. These are associated with what are thought to be freshwater seepage points arising from the lakes Amanzimnyana, nHlange and mPungwini, upstream from Kosi Bay (Fig. 8.5). Seepages from the hardpan dune rock which flow out through the sandy shore are carried northwards by the long shore drift and exactly match the pattern of turtle nesting (Hughes, 1974). Since the loggerheads which breed in Tongaland come from both north and south, any odour cue borne by the freshwater seepage can only be of value in the final stages of nest site selection and of no value in the long-distance navigation of the species.

All the evidence for odour navigation in turtles is circumstantial; no critical experiments have yet been reported which confirm, beyond doubt, the role of olfaction. Nor are there likely to be any for some time, because of the insuperable logistical problems attendant on field studies of navigation behaviour. The possible involvement of visual mechanisms certainly cannot be ruled out, but the difficulty of learning and remembering the position of a tiny island using celestial configurations seems greater than that associated with the recall of odours imprinted in infancy. One certainty remains; if turtles in Brazil can actually perceive and respond to odours emanating from a point source 2250 km away, and use them in their navigation, it will indicate that the sensitivity of their

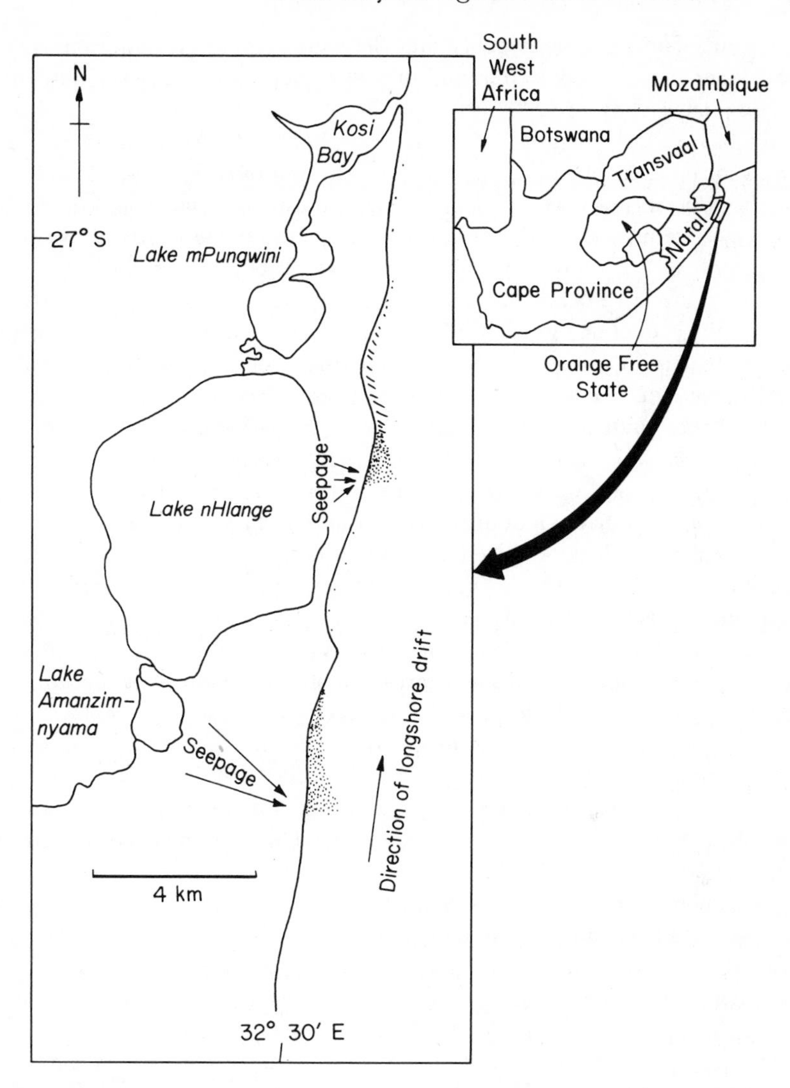

Figure 8.5 Sketch map to show the relation between breeding sites of loggerhead turtles in Tongaland, Natal, and freshwater seepage from the Kosi Bay lakes system. Density of stippling denotes relative density of turtle breeding. (Modified from Hughes, 1974.)

olfactory system easily equals, if not surpasses, that of the fishes.

The only other class of vertebrates which exhibits long-distance breeding migrations is the birds. As far as is known, navigation depends upon eyesight, some species using the sun, some the stars, and others both, or on a gravitational perception. There is some evidence that olfaction is employed, as in fishes, in the final stages of nest site selection and in the correct homing during subsequent foraging or feeding trips. A few species of birds commonly fly at night, particularly those belonging to the order Procellariiforms. Most shearwaters and petrels locate their breeding burrows at night and often in conditions of poor visibility due to low cloud or fog. Procellariiform birds have well-developed olfactory systems, with a high index of olfactory bulb size to size of forebrain, second only to that of the kiwi (Bang and Cobb, 1968), so it is quite likely that olfaction is used in a behaviour which occurs when vision can be of limited use. Besides, the eyes of one procellariiform, the Manx shearwater, *Puffinus puffinus*, are poorly dark-adaptable, suggesting that vision does not play a major role in night navigation (Lockie, 1952). There is some circumstantial evidence that olfactory cues guide incoming Leach's petrel, *Oceanodroma leucorrhoa*, and snow petrels, *Pagodroma nivea*, to their burrows. Birds always land a few metres from the burrow entrance and walk upwind to the entrance. In conditions of still air the approach to the burrow mouth is from a random direction. Significantly, the birds often swing their outstretched heads from side to side in front of them, describing arcs of from 45° to 90°, as if in an attempt to create an asymmetry of perception in each nasal cavity so allowing an accurate klinotactic assessment of the direction of the odour source to be made. A feature of landing flights of these birds is that trees and other obstacles are often blundered into, corroborating the idea that vision plays little part. In an attempt to test the hypothesis that olfaction plays a dominant role in burrow detection by homing petrels, two experimental procedures were employed (Grubb, 1974). Plasticine plugs were pressed into the external nares of one group of test subjects with a batch of controls undergoing identical handling trauma. 93.5% of the controls returned to their burrow within five days, while none of the birds with plugged nares returned home. Plugging the nares interferes with the physiological process of salt gland drainage, however, and the failure to return home might be the result of this disturbance. In a second experiment designed to disrupt olfaction without influencing salt elimination, the olfactory nerves of a group of petrels were transected immediately anterior to the olfactory bulb. A set of the same size underwent operation trauma but the nerves were not transected. By the end of the eighth day after operation 74% of the sham-operated controls had returned but none of the experimental birds had appeared. The interpretation of these data is not clear. Circumstantially it appears that loss of the olfactory ability results in loss of the ability to find home. As was mentioned earlier (p. 79), there is evidence that procellariiform birds locate their fish food by olfaction, so in the experiment outlined above the treated birds may have died of starvation. Nevertheless the results of the plugging experiments are very significant, especially since it has been reported that homing behaviour in the noddy tern, *Anous stolidus*, another species with a salt elimination problem, is unaffected by the application of narial plugs (Watson, 1910). The specific odour cue is not known, but most procellariiform birds squirt stomach oil around the nest burrow entrance during territorial disputes (Jouventin, 1977). In some areas the oily deposit builds up to a thickness of 20 mm or more and retains its pungent and fetid odour. Such a deposition seems likely to act as the odour beacon.

The problem of homing behaviour in the pigeon, *Columba columba*, has aroused the fascination of zoologists for almost two centuries. Pigeons have homing abilities every bit as good as that seen by true migrant species, and return to their home loft from up to 1500 km distance in a time which precludes any random searching for the right way at any stage of the journey. Probably correct navigation depends upon the integrated utilization of several senses; sun compass and, more recently, magnetic wave navigation hypotheses have much supporting evidence. But so does a hypothesis based on olfactory navigation. The hypothesis is based on observations of birds which have had their olfactory nerves cut and others which have had small plastic plugs inserted into the narial apertures. Treated pigeons are far less successful in homing, i.e., they take longer and more of them fail to return all together, than untreated controls. As is shown in Fig. 8.6, treated birds return home more effectively from familiar sites than from unfamiliar sites, indicating that olfactory deprivation does not eliminate homing behaviour and visually recognizable cues can provide sufficient intelligence for an olfactorily deprived bird. The sample size of birds subjected to narial blockage is small and the results must be interpreted with caution, but there is a strong indication from these data of the importance of the olfactory sense. Proponents of the olfactory hypothesis argue that the olfactory quality, encountered on the outward journey towards the point of release, influences the directness of the return flight and that the direction to be taken upon release is fixed by perception of either the odours of home, if the wind conditions and distances allow, or the odours of areas passed *en route* (Papi, 1976). Perception of these known odours results in the setting of the sun or magnetic compass upon which the pigeon relies until it reaches the immediate environs of its home roost. Although it is widely thought that the final stage of roost site selection is governed by the bird's memory of visual landmarks surrounding and adjacent to the loft, olfaction may play a role here too. Swifts, *Apus apus*, deprived of olfactory perception and removed from their nests, are unable to relocate them (Fiaschi *et al.*, 1974), suggesting that olfaction is not totally unimportant, although such studies as have been performed have not demonstrated unequivocally that olfaction plays either a vital role or anything other than only part of an integrated home-finding sensory system.

8.2. Homing orientation

Homing orientation may be defined as an alteration in the position or stance of an animal in response to a sensory stimulus emanating from the home area. Orientation stimuli are acoustic, visual, olfactory and kinaesthetic, and any one

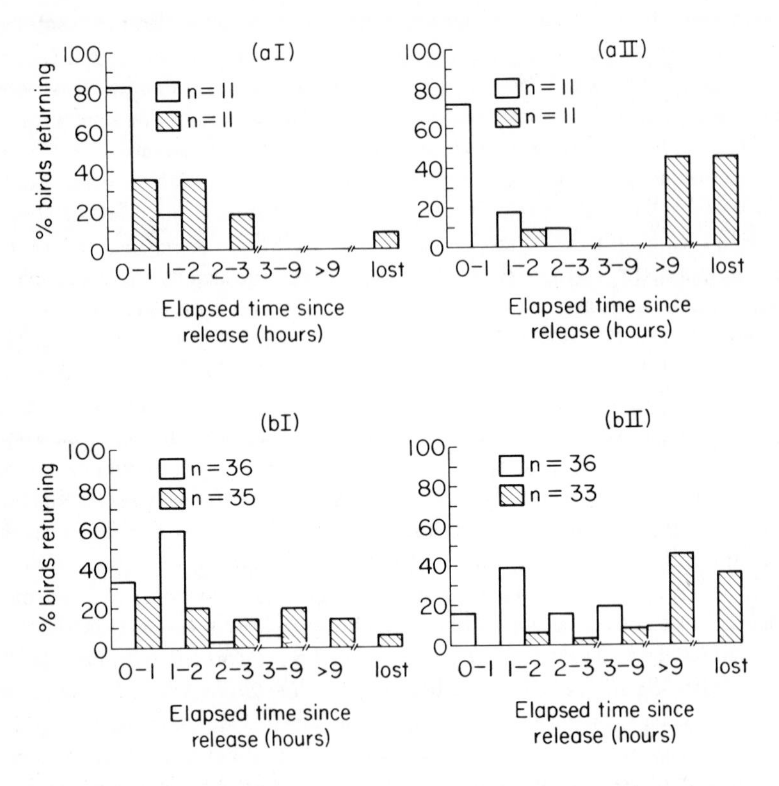

Figure 8.6 Homing performance of anosmic and control homing pigeons. (aI) and (aII) [hatched] Treated pigeons subject to olfactory nerve transection; [open] controls. aI birds released from familiar site 43 km from home. aII birds released from unfamiliar site 34 km from home.
(bI) and (bII) Treated pigeons [hatched] subject to narial blockage; [open] controls. bI birds released from familiar site 55 km from home. bII birds released from unfamiliar site 48 km from home. (Constructed from data in Papi, 1976.)

orientation movement may be affected by a composite consisting of more than one stimulus type. For this reason it is extremely difficult for sensory physiologists to conduct adequate deprivation experiments. Although this book is about olfaction, the reader will have noted many descriptions of behavioural events in which vision, hearing or the kinaesthetic sense play a large role. Nowhere is this problem more sharply defined than in the analysis of homing orientation by olfactory cues, for not only must the olfactory capability of the species be equal to the task demanded of it, but the most rigorous of experiments must be performed to exclude the influence of the other senses. All too frequently experimental rigour is seen to be lacking and the conclusions may not be fully justified. An example may help to make the difficulty clear.

Amphibians have well-developed olfactory systems and, as noted earlier there is good evidence that odour cues lead some species to their invertebrate prey. When displaced from their normal home range, amphibians show a good ability to return. The newt *Taricha rivularis* makes a direct apparently orientated journey to its home area from as far as 0.8 km, and makes a correct initial homeward heading from many times that distance (Twitty, Grant and Anderson, 1966, 1967). Blinding newts prior to their displacement does not adversely affect their ability to return home, but it must be stated that the possible existence and importance of a non-optic photosensory system has not been investigated. By contrast, surgical interruption of the olfactory nerves causes a total loss of homing ability following displacement. However, when a group of control newts is subjected to a sham operation, both these and anosmic experimentals show a random orientation at the site of release (Grant, Anderson and Twitty, 1968). Studies such as these illustrate that the trauma induced by neural mutilation techniques may obscure the basic biological phenomena, and experienced workers with amphibia report that great changes in newt behaviour are wrought by experimental handling. Perfusion of the olfactory cavity with formaldehyde, which induces necrosis in the olfactory membrane, results in a significant loss of orientation ability when treated newts are subsequently released. Animals which have had only their oral cavities perfused with formaldehyde show no impairment of ability. It is important to note that the nasal cavity perfused individuals show only a significant and not a total loss of orientation ability. Possibly non-olfactory stimuli are responsible for some treated individuals to home normally; possibly nasal perfusion does not induce total anosmia; or possibly anosmic newts suffer impaired feeding ability which itself exerts an influence on homing behaviour. *Taricha rivularis* does not appear to be unique with respect to its homing ability. Blinded specimens of the red-cheeked salamander, *Plethodon jordani*, show no impairment of homing ability, (but once again non-optic photosensory systems have not been considered), and surgical mutilation of the nasal area brings about a very significant impairment of ability (Madison, 1969). However, when the olfactorily deprived animals were subsequently subjected to autopsy, it was found that the surgical technique had succeeded only to separate the anterior two-thirds of the olfactory epitheluim, leaving the posterior one-third directly innervated by the olfactory nerve. Since most sensory systems have excess capacity, it is likely that it was the operation trauma, and not the loss of the olfactory sense, which brought about the impairment of homing orientation.

Looking at the problem from a different viewpoint, electrophysiological and anatomical studies on the olfactory membrane of frogs indicate a well-functioning olfactory sense. When presented with a complex mixture of aromatic compounds or some single aromatic compounds known to be attractive, frogs, *Rana temporaria*, move towards the source of the odour diffusion. In a carefully controlled series of experiments, frogs made very significant movements towards the odours, even though they made no selective choice towards either any particular part of their test arena or to any physical feature of their darkened laboratory (Müller and Kiepenheuer, 1976). Further evidence that their olfactory system has sufficient capability to support orientation comes from experiments in which adult Mexican toads, *Bufo valliceps*, are trained to follow visual and olfactory cues in a maze. Although the odorous stimuli are readily available commercial odorants such as cedarwood oil, citral, geraniol, limonene, vanillan, etc., toads learn to recognize their olfactory characteristics and use their memory of them when navigating a maze without the added stimulus of a goal (Grubb, 1976). Olfactometry aids in establishing the existence of a viable orientation system. The toad, *Bufo woodhousei*, and the frogs, *Pseudacris clarki* and *P. streckeri*, orientate repeatedly to a sample of their home pond water and ignore a sample from another pond. However the tree, frogs, *Hyla crysoscelis* and *Scapliopus hurteri*, show no special interest in a sample taken from a pond in which others have spawned, but these frogs breed in ephemeral water bodies and are not attached to a traditional pond like the majority of land anurans. *Bufo woodhousei* is able to find its breeding pond by visual navigation, so this olfactometer revealed ability emphasizes that dependence is not upon only one sensory channel (Grubb, 1973).

If the system is capable of supporting orientation, can unequivocal evidence that it actually does so be obtained from field studies? So far the answer would appear to be no. In a field experiment using leopard frogs, *Rana pipiens*, anosmia was induced by making a scissors cut upwards through the parasphenoid bones. Autopsy showed this practice to be completely satisfactory. Frogs were blinded by severance of the optic nerve behind the eye and deafened by removal of the tympanic membrane and columella. Eyeless frogs return with unimpaired ability to their home site irrespective of whether their point of release is upwind or downwind of their home. Anosmic frogs orientate as well as intact individuals and so do deaf frogs. Even in conditions of heavy fog, when visibility is reduced to a few inches, anosmic frogs orientate unerringly, indicating that neither olfaction nor vision is *essential* for direction finding (Dole, 1972). The interpretation of these results must be that, while the olfactory system alone may be shown to be adequate for orientation behaviour, in the field information is received from many sensory channels which together far exceed the capacity of any single sensory system.

It can be seen from the above illustration that it is extremely difficult to determine with any precision the extent of the role of olfaction in homing orientation. The examples chosen relate to amphibians, but the argument applies equally to all classes of vertebrates. For some curious reason the easily demonstrable homing ability of mammals has not been subjected to the same amount of close scrutiny as the homing ability of amphibians following displacement. The widely known instances of cats and dogs returning to a former home after house removal have never been adequately researched, but there can be little doubt that their navigational ability is based on a complex of sensory interactions. Orientation within a home range or territory, with respect to food, water, nest site, conspecifics, heterospecifics, etc., represents, over short distances and with several different types of sensory stimulus, the same principles that apply to orientation when the animal is displaced from its normal range.

8.3 Summary

There is evidence that anadromous fishes select their natal headstream for the deposition of their own eggs by responding to olfactory cues present in the water. These cues are organic in nature and originate either in the soil or vegetation of the headstream catchment or in the mucus secretions of conspecifics. Oceanic navigation appears to depend upon celestial and solar features. Because of the likelihood that a visual navigation system cannot operate in turtles which breed far from the feeding grounds, an hypothesis of olfactory navigation is attractive. It has yet to be subjected to rigorous field experimentation. The homing ability of pigeons, which is derived from the normal foraging behaviour in rock doves, appears to depend, in part at least, upon the recognition of olfactory landmarks, either learned from exposure to them at the home loft or encountered in the displacement journey. Pigeons rendered olfactorily inadequate are slower in returning to the loft than unimpaired controls, and very many more never return.

Artificial displacement studies with newts and frogs indicate some of the problems involved in assessing the importance of one sensory system in an integrated control system. Although the olfactory system of frogs has been shown to be equal to the task of supporting homing orientation, field studies using olfactorily deprived individuals have failed to demonstrate an elimination of homing behaviour. This failure underlines the complexity of the sensory control mechanisms in behaviours which return animals to their home sites or to their breeding ranges and, by implication, the selective premium placed on such behaviours.

9

Applications of researches into olfactory biology

If ecological research has any final goal towards which it is reaching, that goal must concern the quality of the natural environment and man's place in it. More specifically, it embraces man's problems and difficulties in raising more crops from ecosystems continuously striving to change, and in keeping those crops free from debilitating pest infestations. Such goals are an extension of those reached for in many areas of biological research and which concern the physiological and behavioural functioning of single individuals or defined social groups. To suggest that studies in olfactory biology have progressed to the point where their findings can be applied to the field and can be shown to take us nearer to that goal would not be as true for vertebrate animals as it is for invertebrates. Since about 1964, when insect sex attractant pheromones were first identified and synthesized, olfactory biology has played an escalating role in the formulation of pest control programmes. Sex pheromone traps are now used in control measures against many insect pests worldwide with encouraging results. In the United States alone the cotton bollweevil, *Anthonomus grandis*, is estimated to depress the value of cotton production by $300 million annually. One-third of all insecticides used in the U.S.A. for agricultural purposes, and costing $70 million annually, is used in the battle against the weevil. The synthetic sex pheromone, called 'grandlure' after the specific name of the weevil, is now one of the most promising pheromones being used in the detection, supression and control of any insect species. The cabbage looper, *Trichoplusia ni*; codling moth, *Laspeyresia pomonella*; red-banded leaf-roller, *Argyrotaenia velutinana*; pink bullworm of cotton, *Pectinophora gossypiella*; gypsy moth *Porthetria dispar*; spruce budworm, *Choristoneura fumiferana*; and pine bark beetle, *Dendrochtonus brevicomis* are among a host of other insect species of economic importance which are subject to intense pheromone control studies (for a comprehensive review of this field see Birch *et al.*, 1974; Trammel, 1976).

Part of the reason for this phenomenal success is that insects respond blindly to the messages carried in sex pheromones. For vertebrates this is not so and factors such as social dominance status, amount of previous exposure to the signal, age, nutritional status, and ability to communicate visually and acoustically at the same time as olfactorily conspire to frustrate simple olfactory attempts at controlling behaviour. But not all the signs are gloomy, and in some areas the judicial broadcast of appropriate odours at the right time is revealing encouraging trends. Because of the complexity of vertebrate behaviour and its mesh of releasers, application of odours is unlikely ever to be as effective as it is for insects and other invertebrates. With the losses of stored products to rodent pests running at a conservative estimate of some hundreds of millions of dollars annually, any partially effective control may prove economically most worthwhile.

This chapter will attempt to review the advances made in olfactory pest control and in the other area where advances have been made, animal husbandry, and to point out where future research emphasis might be most productive. Mankind is rapidly approaching the point at which he has to run to stand still – gains from hardly-won agricultural advances are quickly being dispelled by pests which seem more difficult to dislodge as more means are found to control them with traditional techniques. The time has come for a more adventurous approach to the problem, if we are not soon to start sprinting.

9.1 Pest repellents and attractants

There are two ways in which vertebrate odours can be thought of as acting, either directly bringing about an immediate change in behaviour or indirectly through endocrine activity on the central nervous system (Wilson and Bossert, 1963). Into this latter category fall the pheromones known to block pregnancy and otherwise influence reproductive physiology. If they had a guaranteed 100% effect they would indeed have great potential application in the field of rodent pest control. Unfortunately not all females succumb to pregnancy block on exposure to strange male urine and the proportion can be as low as 70 %. The reproductive prowess of rodents being what it is, anything less than a 100% mortality is ineffective as a control, as a replacement population can soon grow from a handful of survivors (Drummond, 1970; Kenelly *et al.*, 1972). The most fruitful area for research and in which the greatest strides have been made is with the direct acting, or signalling, odours. Short-term relief from a pest problem can be obtained by repelling pests from the crop at risk – this might occur if the crop was only attractive to the pest during a short period, perhaps immediately prior to harvest. For stored products, or crops that retain their attractiveness to pests over a long period of time, repulsion requires a sustained effort and probably necessitates repeated applications of ever more powerful deterrents. To take a simple analogy, a crude scarecrow in a field has a dramatic effect for a short while following its erection, but soon after it serves as a convenient perch for those it was designed to frighten. In such cases the only effective way forward is to impose mortality through whatever means possible. For rodents, this means attracting them to baiting stations where they can eat a favoured food spiked with anticoagulants, chemosterilants, or acute rodenticides. Rats are notoriously suspicious of bait piles and can detect human odours associated with the baiting trays. Clearly, any increase in the number of rats attracted to the bait, or any increase in the amount of bait consumed, will result in a higher mortality rate. Intense study of the biology of the ricefield rat, *Rattus mindanensis*, has revealed that this can be achieved using an attractive food odour (Shumake, 1977). Zinc phosphide is a widely used acute rodenticide, but if a rat consumes a small dose, not only will it survive, but it will not touch poisoned bait again. It is therefore important that the poison be mixed with a highly attractive bait base. Table 9.1 shows the amount of rice eaten by groups of five *R. mindanensis* over a period of eight days, and it can clearly be seen that the addition of trapped rice volatiles to granulated grain very significantly increases its palatability. When zinc phosphide is added to granulated rice treated with rice volatiles and to untreated ground rice, the uptake of the poison is almost twice as high in those rats feeding on the treated bait (33.9 mg/kg rat/day *vs.* 19.8 mg/kg rat/day). After three days' exposure to the poison and a fortnight follow-up observation period, fourteen of the treated bait feeding rats were dead while only eight of those feeding on plain

Table 9.1 The influence of added rice volatiles to the amount of rice eaten each day by four groups each of five *Rattus minandensis*. The weight of rice eaten, in grams, is shown in the respective columns (Means ± SE). (From Shumake, 1977.)

	Whole grain rice	Ground rice	Ground rice treated with 1% soybean oil	Ground rice treated with trapped rice volatiles
1	8.1 ± 1.7	12.9 ± 2.7	22.8 ± 2.9	35.1 ± 3.2 †
2	6.6 ± 2.1	8.9 ± 1.6	21.9 ± 2.3	35.1 ± 3.2 *
3	5.3 ± 2.0	15.5 ± 2.7	25.9 ± 2.5	35.9 ± 3.3 †
4	6.4 ± 2.4	14.6 ± 2.9	16.4 ± 3.0	34.4 ± 2.8 †
5	5.8 ± 2.0	12.0 ± 2.3	19.8 ± 3.0	34.6 ± 2.6 †
6	4.5 ± 1.7	15.6 ± 2.6	21.6 ± 2.6	36.8 ± 3.1 ‡
7	5.5 ± 2.6	18.9 ± 2.4	15.0 ± 2.2	35.8 ± 2.7 †
8	7.0 ± 2.8	19.1 ± 3.3	21.9 ± 3.0	32.4 ± 1.7 †

Probability that amount of treated rice eaten exceeds that of whole grain rice:
* = < 0.10
† = < 0.05
‡ = < 0.01

bait had died. While these results are not statistically significant, they do indicate an encouraging trend which should be followed up with the minimum of delay.

Few attempts have been made to examine the possibility of using sex attractants present in the urine to attract rodents to traps or other killing sites. Shumake (1977) found that only the urine of female rats in pre-oestrus elicited any behavioural change in male rats, but male rat urine is reported as attracting all females, in studies conducted in the People's Republic of China (Anonymous, 1975). Furthermore it is stated that males can be attracted to female urine, being enticed out of territorial inquisitiveness. These suggestions of behavioural effect are in need of critical reassessment, but the physiological effect of the odour of female urine on the blood testosterone concentration of males is beyond doubt (Purvis and Haynes, 1978).

Part of the problem of scientific pest control is the assessment of the numbers of the pest. Control programmes have to be economically acceptable to all parties concerned and this necessitates accurate scrutiny of numbers. An effective means of doing this is to attract the pest individuals to a small number of assessment points, suitably prepared so that the number or density of tracks or spoor can be assessed over a standard time interval and the index obtained compared with that of previous years. In the western part of the United States, the coyote, *Canis latrans*, remains a major threat to sheep farmers and costs the country million of dollars annually. Since the early 1970s, attractive olfactory lures have been used in this exercise and have revealed that the coyote population is still expanding (Linhart, 1973). The attractant, which is placed in a small capsule and buried just beneath the centre of a patch of raked soil, is a commercially made fermented egg product. The number of tracks in the soft earth is counted each day for a prescribed period. Substances like this, or like the synthetic fermented egg being developed by the Wildlife Research Center in Denver, Colorado, may also be of use in attracting coyotes to trapping or gassing sites, but this possibility is still being examined.

So far the search for an olfactory repellent to shield sheep against coyote attack has been largely unsuccessful, because any substance affecting the coyote also adversely affects the sheep. The most effective is β-chloro-acetyl chloride; in a series of test trials it repelled the advances of coyote and dog alike on every occasion (Lehner *et al.*, 1976). However, this is a strong mucus membrane irritant which induces profuse lacrimation in predator and prey alike, and is only repugnant – not fearful – to coyotes. Efforts to protect crops from the ravages of ungulates through the use of olfactory repellents have met with mixed response. Out of a very large number of substances presented to black-tailed deer, *Odocoileus virginianus*, only three have been found to have repellent properties. They are a fermented egg product, a putrifying fish product, and a proteinaceous material used in rendering animal tissue. Abattoir offal is also effective, but for reasons of hygiene somewhat unacceptable for widespread use. Congealed blood mixed with whale oil has been shown to have some effect in field trials. Grey duiker, *Sylvicapra grimmia*, in South Africa have successfully been kept out of gardens where they were causing damage to beans and avocado pear saplings after plastic bags containing the repellent were hung on the garden fence (Wright and Bourquin, 1977). In further trials, with adequate controls, bags of repellent hung at 15 m or 30 m intervals around a field successfully deterred reedbuck, *Redunca arundinum*, from entering, but had less effect on steenbok, *Raphicerus campestris*, and oribi, *Ourebia ourebi*. However, the promise of these trials was not maintained when reedbuck were causing damage estimated at between $11,000 and $22,000 per night to a field of irrigated lettuce. The repellent was of no use in this instance.

The use of naturally occurring, but ignored plants, has received some attention. Wild ginger, *Asarum candatum*, is common in the northwestern part of the U.S.A. but seldom, if ever, does it occur in the browse of *Cervus canadensis* and *Alces alces*. Captive deer and elk reject their food if ground ginger is mixed with it, and the damage to young Douglas fir seedlings treated with a solution of it is much reduced (Campbell and Bullard, 1972). Further work with distasteful and aromatic plants is urgently needed.

9.2 Animal husbandry

It is necessary for animal flesh producers to adopt husbandry procedures which are as efficient as possible: the maximum amount of meat must be produced for the minimum of time and money expenditure. This means that the conditions under which the stock live hardly begins to resemble that known by their forefathers before domestication. Only species that can adopt to conditions of captivity have been domesticated, but even so there is evidence that productivity performance is lower than it could be. As was shown in Chapter 4, synthetic porcine pheromone is now used widely to aid the pig breeder to detect those sows in oestrus. Furthermore, rebred sows, that is, sows mated immediately after having weaned a litter, come into oestrus significantly sooner if they have been treated with porcine pheromone during the first few days following the birth of their previous litter than if they have been kept isolated from the odour (Hillyer, 1976). The saving in time is of the order of two weeks or more and an additional advantage is that the average weight of the piglets at birth is slightly higher than that of piglets born to untreated sows though the number is the same. All in all the use of just a single one-second-long spray of odour immediately after birth could increase annual productivity appreciably. In Great Britain something like 14 million pigs are slaughtered annually; a 1% increase represents 140 000 pigs with a total market value to the producers of around £5 million (in 1977). Although the odorous environment of other species does not seem to play such an overwhelming role as it does in the pig, veterinary surgeons, zookeepers and others involved with the breeding of endangered species would be well advised to consider how the environment can be improved. Sperm banks stocked with semen from rare or threatened species may be only part of the solution; body odour may be equally important (Hatlapa, 1977).

The value of dressing an orphaned lamb in the skin of a dead lamb so the deprived mother will suckle the orphan is well known and is a routine aspect of hill farming practice. The same technique is used in parts of South America to obtain wool of the vicuna, *Lama vicugna*, in quantities found in the alpaca, *L. pacos*. A newborn male vicuna is dressed in the skin of a dead alpaca and suckled by a female alpaca upon which species the young vicuna becomes sexually imprinted. When it matures it mates with a female alpaca and the resulting hybrid produces vicuna wool in alpaca quantities (Hodge, 1946). The ability shown by ungulates of interbreeding following olfactory imprinting during infancy has a great potential which is as yet quite unexplored.

Olfactory biology has a role to play in fish husbandry, both in the rearing of fishes in hatcheries and fish farms and in the exploitation of natural populations. Fishes appear to require certain scents for the elicitation of feeding behaviour, and although generalized herbivores such as carp can be fed on chopped and fermented sweet corn, *Zea mais* (Phillipson, 1967), the farming of carnivorous species presents great problems. Most carnivorous fishes hunt with their noses and feeding behaviour is elicited only after certain specific scent signals have been received (Kleerekoper and Morgensen, 1963). Developers of fish farms should consider the implications of this very seriously.

Fishes respond to very low concentrations of odorants to guide them home, and to cause them to scatter or to shoal for protective or reproductive purposes. Such behaviours can have commercial application but only very occasionally have olfactory lures actually been used in commercial fisheries; for example, fishermen in the Mississippi river system use caged channel female catfish, *Ictalurus punctatus*, to draw large numbers of males into their nets (Timms and Kleerekoper, 1972). The behavioural factors keeping shoals together are many and complex, but in those species which have been examined specific odorants certainly play a part (Bloom and Perlmutter, 1977). Further studies may show that natural shoal size may be enhanced, and possibly even cleaned of unwanted other fishes, by the applications of the quite small quantities of specific odorants to the fishes' feeding grounds.

In the semi-natural environments of fish cultivation ponds or angling club ponds, excessive stocking may not only depress individual growth rates through competition for the available food, but it may also depress natural reproduction through the operation of a spawning inhibitor substance (Swingle, 1953). Overstocking may breed further overstocking. Solomon (1977) suggests that selective breeding studies may be required to develop strains with reduced inhibitor substance production for situations requiring high stocking rates. Because of the high rates of primary and secondary productivity found in aquatic ecosystems, development and application of the findings of fish olfactory biology to pisciculture and fisheries' techniques may prove of more immediate value to mankind than the application of mammalian olfactory biology to stock husbandry, though in the long term the benefit to mankind from both will be found to be great.

9.3 Miscellaneous applications

It is a feature of human behaviour that the sensory or physical attributes of animals are never adequately researched until mankind requires either to use the natural ability of the species for a specific purpose or to make a machine which incorporates some design features already found in nature. Thus much attention has been paid to the olfactory prowess of dogs because man can use dogs' noses for forensic and tracking purposes and dogs can be domesticated and trained simply. Or again, study of the physics and behaviour of echolocation in bats received a massive shot-in-the-arm following the development of radar for military purposes. It is a pity events should course like this because it restricts the rate at which fundamental, exciting and potentially useful discoveries can be made. For example, although the olfactory powers of dogs cannot be denied, those of fish are very much greater. While it is acknowledged that fish cannot be used for tracking purposes, they are readily trainable and could be used to detect poisons and other substances at concentrations far lower than could dogs. This may have particular application in medical diagnosis where the tiniest quantities of specific metabolites in the urine, too low for present bioassay techniques to detect, could invoke an unequivocal response. In this context it is interesting to note that doctors and paramedics are now being trained to recognize the odours of certain conditions. It is vital for paramedics and other emergency treatment personnel to distinguish, for example, ketoacidosis induced by diabetes from alcoholic or drug overdose when starting treatment on an unconscious patient. In this case the sweet, acetone-like smell of diabetes is very distinctive. Microencapsulated samples of this and other distinctive odours are now used in several medical training programmes (Lukas, Berner and Kanakis, 1977).

One area of interest to the medical profession which has received very little attention as yet is the role of odours in psychotherapy. There is some evidence that the olfactory environment of the young child is a necessary adjunct to the development of correct sexual identity, though the evidence is largely anecdotal and not capable of withstanding rigorous examination. However, the well-researched studies into olfactory imprinting in mammals suggest that early olfactory contact is a normal and necessary first step towards the later establishment of firm social relationships. While it may suit the work routines in some maternity hospitals to separate mothers from their newborn children for up to 24 hours as a matter of convenience, the loss of the initial bond-forming development period that this causes may create untold social wretchedness in later life. Attitudes in this area are changing, thankfully, and the gathering trend back towards breast feeding in developed countries means that an ever-increasing number of infants are allowed to make early olfactory contact with their mothers' mammary glands – organs which abound in specialized sebaceous glands and impart a particular odour to the newborn infant. Extension of studies to include the therapeutic effects of odours and odour mimics may open up huge possibilities for psychological research.

The possible existence of human pheromones has been argued from a few rather anecdotal studies and from extrapolation from studies on animals (Comfort, 1971a, b; McClintock, 1971). There is little controlled experimentation possible to verify their existence because of the artificial environment with which mankind cossets himself, and any role they may have is unlikely to be more than supportive to the other senses. This is no reason to dismiss their potential importance in many aspects of human life and strides should be made by reputable psychological laboratories to arrive at a meaningful assessment of their capabilities. There is no doubt that naturally produced odour is important to man, for from the earliest times of recorded history he has striven to mask his own smell with tinctures made from aromatic animals and plants. In the developed world of today the growth of the male perfume industry is phenomenal and the profits are rapidly catching up with those made in the female perfume industry. It is high time we understood our psychological relationship with body odour.

Turning away from humans, it is hard to see how any further development of olfactory lures can aid the plight of fur trappers. For generations fur trappers have used natural musks of the species they wish to trap to entice inquisitive individuals into their traps. Most of these species are solitary carnivores which use the musk to mark their territorial boundaries – if such substances only said 'keep out' the trappers would long ago have abandoned the technique. Territory boundaries are meant to be visited, so it is likely that the trappers are using the very best olfactory technique now justified by modern research. The fur-trapping industry is hardly a growth industry and may almost disappear over the next few decades as natural stocks of fur-bearing mammals dwindle.

As human beings consume ever-increasing amounts of energy, the need for gas and oil pipelines will continue to grow. Construction firms normally check for leaks in section joints with laser and ultrasonic equipment, but the field observations of a Texan engineer four decades ago indicate that a cheaper means is available in some parts of the world, at least. If a high concentration of ethyl mercaptan is pumped into an unchecked pipeline, some of the evil-smelling compound will escape from the leaks. This is readily detected by turkey vultures, *Cathartes aura*, which will circle around the rising plume of odorant and descend to remain in close proximity to the leak. The only equipment the pipeline engineer needs for the location of the leak is a pair of binoculars and a portable anemometer (Stager, 1964). This technique will only work where there are turkey vultures, but they range widely in the New World from Hudson's Bay in the North (55°N) to Magellan's Straits in the South (52°S).

Coupled with man's insatiable desire for energy is his uncompromising production of industrial waste. The problems faced by anadromous fish in navigating our rivers on their way to their natal headstreams are greater now than ever before. Dams and power races force salmon to use narrow runs but the real danger comes from the changed water compositions resulting from pollutants. Lowered oxygen tension, a common result of eutrophication following poor sewage treatment, seriously affects the salmon's powers of olfactory discrimination (Hasler, 1966). The effect of heavy metals and biochemical pollution on olfaction can only be guessed at. The wise management of salmon stocks now threatened with extinction must be based upon urgently conducted studies into their navigational biology and in particular into their reliance upon olfaction.

Finally, mention should be made of man's attempts to develop an effective shark repellent of most potential use in aircraft pilot's and seamen's survival kits. Sharks can be demonstrated to locate their prey with their noses, so the development of an olfactory repellent seems to be a logical possibility. So far no substance has been found which is invariably successful and, in spite of a great deal of research effort being expended on the food-selecting behaviour of sharks, it is apparent that no simple answer will emerge. In spite of their lowly taxonomic position, sharks are intelligent creatures who may rely upon their noses to a differential extent depending upon their mood and current environmental conditions. If this is so, the search for a universal panacea is destined to failure, just as much as the search for simple olfactory governors of mammalian behaviour is bound to fail for many of the same reasons.

It would be wrong to end this brief discussion on a gloomy note. It is true that the applications of olfactory means of pest control and animal husbandry are few in number and those that are in use are still in their infancy. Perhaps workers on vertebrate species are daunted by the immediacy of the successes scored by their colleagues working on invertebrates. There are probably many areas in which they could work together, although so far only one has been reported. However, it is an area of some significance. Klipspringer antelopes, *O. oreotragus*, use the secretion of their pre-orbital glands to mark grasses and twigs in their ranges, particularly at the boundaries. Rainfall leaches an as yet unidentified ingredient out of the resinous mark which runs down the stem to the ground and which is highly attractive to adult ticks of the species *Ixodes neitzi*. Following the course of this leachate the ticks gather at the resinous mark in quite large numbers waiting to jump onto the next klipspringer which pauses to sniff the mark (Rechav *et al.*, 1978). Here, knowledge of the scent-marking behaviour of the antelope can be combined with the host-finding behaviour of the parasite to produce a meaningful and effective control programme for the tick. Future research may reveal this type of mechanism not to be unique to the *Oreotragus – I. neitzi* interaction but to be quite widespread. If so, it may open up new horizons for research into economically important arthropod ectoparasites of large mammals.

As ecologists concerned with population and pest control become more aware of the behavioural interactions between the animal, its conspecifics and its environment, the amount of research effort directed at olfactory phenomena will surely increase. The application of this knowledge will benefit mankind by not increasing the amount of toxic and cumulative debilitating substances in the environment, yet protecting his growing or stored crops and stock. Lessons learned from practical control programmes indicate that selective killing with the snare, trap, poisoned bait or gun do little more than bring a temporary solution. The use of substances which continually alter behaviour – perhaps inducing dispersal or preventing mating – offers a more attractive solution, even if it is, at present, just beyond our reach.

References

Adams, J., Garcia, A. and Foote, C. S. (1968), Some chemical constituents of the secretion from the temporal gland of the African elephant (*Loxodonta africana*). *J. chem. Ecol.*, **4**, 17–25.

Albone, E. S. (1975), Dihydroactinidiolide in the supracaudal gland secretion of the red fox. *Nature*, **256**, 575.

Albone, E. S. and Eglinton, G. (1974), The anal sac secretion of the red fox (*Vulpes vulpes*); its chemistry and microbiology. A comparison with the anal sac secretion of the lion (*Panthera leo*). *Life Sci.*, **14**, 387–400.

Albone, E. S. and Perry, G. L. (1976), Anal sac secretion of the red fox, *Vulpes vulpes*; volatile fatty acids and amines: implications for a fermentation hypothesis. *J. chem. Ecol.*, **2**, 101–11.

Aleksiuk, M. (1968), Scent-mound communication, territoriality, and population regulation in beaver (*Castor canadensis* Kuhl.).

Allee, W. C., Finkel, A. J. and Hoskins, W. H. (1940), The growth of goldfish in homotypically conditioned water; a population study in mass physiology. *J. exp. Zool.*, **84**, 417–43.

Allen, G. M. (1939), *Bats*, Harvard University Press, Cambridge, Mass.

Allison, A. C. (1953), The morphology of the olfactory system in the vertebrates. *Biol. Rev.*, **28**, 195–244.

Allison, A. C. and Warwick, R. T. T. (1949), Quantitative observations on the olfactory system of the rabbit. *Brain*, **72**, 186–96.

Andersen, K. K. and Bernstein, D. T. (1975), Some chemical constituents of the scent of the striped skunk (*Mephitis mephitis*). *J. chem. Ecol.*, **1**, 493–501.

Andersson, G., Andersson, K. Brundin, A. and Rappe, C. (1975), Volatile compounds from the tarsal scent gland of reindeer (*Rangifer tarandus*). *J. chem. Ecol.*, **1**, 275–81.

Andrew, R. J. (1963), The origin and evolution of the calls and facial expressions of the primates. *Behaviour*, **20**, 1–109.

Anonymous (1975), The attractive effect of rat urine odour and its possible use in rat control. *Acta zool. sinica*, **21**, 46–50.

Apfelbach, R. (1973), Olfactory sign stimulus for prey selection in polecats (*Putorius putorius* L.). *Z. Tierpsychologie*, **33**, 270–3.

Auffenberg, W. (1965), Sex and species discrimination in two South American tortoises. *Copeia*, 335–42.

Auffenberg, W. (1966), On the courtship of *Gopherus polyphemus. Herpetologica*, **22**, 113–17.

Auffenberg, W. (1969), *Tortoise behavior and survival*, Rand McNally & Co., Chicago, Ill.

Baker, J. R. (1974), *Race*, Oxford University Press, Oxford.

Baldwin, B. A. and Shillito, E. E. (1974), The effects of ablation of the olfactory bulbs on parturition and maternal behaviour in Soay sheep. *Animal Behaviour*, **22**, 220–3.

Bang, B. G. (1960), Anatomical evidence for olfactory function in some species of birds. *Nature*, **188**, 547–9.

Bang, B. G. and Cobb, S. (1968), The size of the olfactory bulb in 108 species of birds. *Auk*, **85**, 55–61.

Bardach, J. E. and Todd, J. H. (1970), Chemical communication in fish, In: *Advances in chemoreception*, I, *Communication by chemical signals* (ed. J. W. Johnston, D. G. Moulton and A. Turk), Appleton-Century-Crofts, New York, pp. 205–40.

Bardach, J. E. and Villars, T. (1974), The chemical senses of fishes, In: *Chemoreception in marine organisms* (ed. P. T. Grant and A. M. Mackie), Academic Press, London, pp. 49–104.

Barnett, S. A. (1963), *A study in behaviour*, Methuen & Co., London.

Barrette, C. (1977), Scent marking in captive muntjacs, *Muntiacus reevesi. Animal Behaviour*, **25**, 536–41.

Baumann, F. (1927), Experimente über den Geruchissinn der Viper. *Rev. suisse zool.*, **34**, 173–84.

Baumann, F. (1928), Über die Bedeutung des bisses und des Geruchissinnes für den Nahrungserwerb der Viper. *Rev. suisse zool.*, **35**, 233–9.

Baylock, L. A., Ruibal, R. and Pratt-Aloia, K. (1976), Skin structure and wiping behavior of phyllomedusine frogs. *Copeia*, 283–95.

Bartholomew, G. A. (1959), Mother-young relations and the maturation of pup behaviour in the Alaska fur seal. *Animal Behaviour*, **7**, 163–71.

Beach, F. A. and Gilmore, R. W. (1949), Response of male dogs to urine from females in heat. *J. Mammal.*, **30**, 391–2.

Beach, F. A. and Jaynes, J. (1956), Studies on maternal retrieving in rats. III. Sensory cues involved in the lactating female's response to her young. *Behaviour*, **10**, 104–25.

Beauchamp, G. K. (1976), Diet influences attractiveness of urine in guinea pigs. *Nature*, **263**, 587–9.

Beauchamp, G. K., Magnus, J. G. Shmunes, N. T. and Durham, T. (1977), Effects of olfactory bulbectomy on social behavior of male guinea pigs (*Cavia porcellus*). *J. comp. physiol. Psychol.*, **91**, 336–46.

Békésy, G. von (1964), Olfactory analogue to directional learning. *J. appl. Physiol.*, **19**, 369–73.

Bennion, R. S. and Parker, W. S. (1976), Field observations on courtship and aggressive behavior in desert striped whipsnakes, *Masticophis t. taeniatus. Herpetologica*, **32**, 30–5.

Bertman, G. and Toft, R. (1969), Sensory mechanisms of homing in salmonid fish. I. Introductory experiments on the olfactory sense in grilse of Baltic Salmon. *Behaviour*, **35**, 235–41.

Berwein, M. (1941), Beobachtungen und Versuche über das gesellige Leben der Ellritze. *Z. vergleichende Physiol.*, **28**, 402–20.

Bhatnagar, K. P. and Kallen, F. C. (1975), Quantitative observation on the nasal epithelic and olfactory innervation in bats. *Acta anatomica*, **91**, 272–82.

Bigelow, H. B. and Schroeder, W. C. (1948), Fishes of the western North Atlantic. III. Sharks. *Memoirs of the Sears Foundation for Marine Research*, **1**, 59–576. New Haven, Conn.

Birch, M. C., Trammel, K., Shorey, H. H., Gaston, L. K., Hardee, D. D., Cameron, E. A., Sanders, C. J., Bedard, W. D., Wood, D. L., Burkholder, W. E. and Müller-Schwarze D. Programs utilizing pheromones in survey or control, In: *Pheromones* (ed. M. C. Birch), North Holland Publishing Co., Amsterdam, 411–61.

Bloom, H. D. and Perlmutter, A. (1977), A sexual aggregating pheromone system in the zebra fish *Brachydanio rerio* (Hamilton-Buchanan). *J. exp. Zool.*, **199**, 215–26.

Bogert, C. M. (1941), Sensory cues used by rattlesnakes in their recognition of ophidian enemies. *Ann. N.Y. Acad. Sci.*, **41**, 329–43.

Bojsen-Møller, F. (1975), Demonstration of terminalis, olfactory, trigeminal and perivascular nerves in the rat nasal septum. *J. comp. Neurol.*, **159**, 245–56.

Bojsen-Møller, F. and Fahrenkrug, J. (1971), Nasal swell-bodies and cyclic changes in the air passage of the rat and rabbit nose. *J. Anat.*, **110**, 25–37.

Bolles, R. C. and Woods, P. J. (1964), The ontogany of behaviour in the albino rat. *Animal Behaviour*, **12**, 427–41.

Boonstra, R. and Krebs, C. J. (1976), The effect of odour on trap response in *Microtus townsendii*. *J. Zool.*, **180**, 467–76.

Bourlière, F. (1955), *The natural history of mammals*, Harrap & Co., London.

Bowers, J. M. and Alexander, B. K. (1967), Mice: Individual recognition by olfactory cues. *Science*, **158**, 1208–10.

Brisbin, I. L. Jr. (1968), Evidence for the use of postanal musk as an alarm device in the King snake, *Lampropeltis getulus*. *Herpetologica*, **24**, 169–70.

Brockie, R. (1976), Self-anointing by wild hedgehogs, *Erinaceus europaeus* in New Zealand. *Animal Behaviour*, **24**, 68–71.

Bronson, F. H. (1971), Rodent pheromones. *Biology of Reproduction*, **4**, 344–57.

Brooksbank, B. W. L., Brown R. and Gustafsson R. (1974), The detection of 5 -Androst-16-en-3 α-ol in human male axillary sweat. *Experientia*, **30**, 864–5.

Brownlee, R. G., Silverstein, R. M., Müller-Schwarze, D., and Singer, A. G. (1969), Isolation, identification and function of the chief component of the male tarsal scent in black-tailed deer. *Nature*, **221**, 284–5.

Bruce, H. M. (1959), An exteroceptive block to pregnancy in the mouse. *Nature*, **184**, 105.

Bruce, H. M. and Parrott, D. M. V. (1960), Role of olfactory sense in pregnancy block by strange males. *Science*, **131**, 1526.

Brundin, A., Andersson, G., Andersson, K., Mossing, T. and Kallquist, L. (1978), Short-chain aliphatic acids in the interdigital gland secretion of reindeer (*Rangifer tarandus* L.), and their discrimination by reindeer. *J. chem. Ecol.*, **4**, 613–22.

Buckland, F. (1880), *Natural history of British fishes*, Unwin, London.

Burger, B. V., Le Roux, M., Garbes, C. F., Spies, H. S., Bigalke, R. C., Pachler, K. G., Wessels, P. L., Christ, V. and Maurer, K. H. (1977), Further compounds from the pedal gland of the bontebok (*Damaliscus dorcas dorcas*). *Z. Naturforsch.*, **32**, 49–56.

Burger, B. V., Le Roux, M., Garbers, C. F., Spies, H. S. C., Bigalke, R., Pachler, K. G. R., Wessels, P. L., Christ, V., Maurer, K. H. (1976), Studies on mammalian pheromones. I. Ketones from the pedal gland of the bontebok (*Damaliscus dorcas dorcas*). *Z. Naturforsch.*, **31**, 21–8.

Burghardt, G. M. (1967), Chemical-cue preferences in inexperienced snakes: Comparative aspects. *Science*, **157**, 718–21.

Burghardt, G. M. (1970), Chemical perception in reptiles, In: *Advances in chemoreception*, I, *Communication by chemical signals*, (ed. J. W. Johnston, D. G. Moulton and A. Turk), Appleton-Century-Crofts, New York, pp. 241–309.

Burkholder, G. L. and Tanner, W. W. (1974), A new gland in *Sceloporus graciosus* males. *Herpetologica*, **22**, 199–206.

Campbell, D. L. and Bullard, R. W. (1972), A preference testing system for evaluating repellents for black-tailed deer. *Proc. Fifth Vertebrate Pest Conf., California*, 56–63.

Carr, A. (1965), The navigation of the green turtle. *Scientific American*, **212**, 78–86.

Carr, W. J., Martorano, R. D. and Krames, L. (1970), Responses of mice to odors associated with stress. *J. comp. physiol. Psychol.*, **71**, 228–38.

Carr, W. J., Loeb, L. S. and Dissinger, M. L. (1965), Responses of rats to sex odors. *J. comp. physiol.* Psychol., **59**, 370–77.

Carter, S. C. and Marr, J. N. (1970), Olfactory imprinting and age variables in the guinea pig, *Cavia porcellus*. *Animal Behaviour*, **18**, 238–244.

Charles-Dominique, P. (1977), Urine marking and territoriality in *Galago alleni* (Waterhouse, 1837-Lorisidae, Primates) – a field study by radio-telemetry. *Z. Tierpsychologie*, **43**, 113–38.

Chiszar, D., Scudder, K. and Knight, L. (1976), Rate of tongue flicking by garter snakes and rattlesnakes during prolonged exposure to food odours. *Behav. Biol.*, **18**, 273–83.

Clulow, F. V. and Clarke, J. R. (1964), Pregnancy block in *Microtus agrestis*, an induced ovulator. *Nature*, **219**, 511.

Coblentz, B. E. (1976), Functions of scent-urination in ungulates with special reference to feral goats (*Capra hircus* L.) *Am. Nat.*, **110**, 549–57.

Cohen, M. and Gerneke, W. H. (1976), Preliminary report on the intermandibular cutaneous glandular area and the infraorbital gland of the steenbok. *J. S. A. Vet. Ass.*, **47**, 35–7.

Colby, D. R. and Vandenberg, J. G. (1974), Regulatory effects of urinary pheromones on puberty in the mouse. *Biology of Reproduction*, **II**, 268–79.

Cole, C. J. (1966), Femoral glands of the lizard, *Crotophythus collaris*, *J. Morphol.*, **118**, 119–136.

Comfort, A. (1971a), Likelihood of human pheromones. *Nature*, **230**, 432–3.

Comfort, A. (1971b), Communication may be odorous. *New Scientist*, Feb. 25, 412–14.

Courtney, R. J., Reid, L. D. and Wasden, R. E. (1968), Suppression of running times by olfactory stimuli. *Psychonomic Sci.*, **12**, 315–6.

Cowles, R. B. (1938), Unusual defense postures assumed by rattlesnakes. *Copeia*, 13–16.

Cowles, R. B. and Phelan, R. L. (1958), Olfaction in rattlesnakes. *Copeia*, 77–83.

Cowley, J. J., Wise, D. R. (1970), Pheromones, growth and behaviour. *CIBA Foundation Study Group*, **35**, 144–70.

Dagg, A. I. and Taub, A. (1970), Flehmen. *Mammalia*, **34**, 686–95.

Dagg, A. I. and Windsor, D. E. (1971), Olfactory discrimination limits in gerbils. *Can. J. Zool.*, **49**, 283–5.

Davies, D. J. and Bellamy, D. (1972), The olfactory response of mice to urine and effects of gonadectomy. *J. Endocrinol.*, **55**, 11–20.

Davies, J. T. and Taylor, F. H. (1958), Molecular shape, size and absorption in olfaction. *Proc. 2nd Int. Cong. Surface Activity*, **4**, 329–40.

Davis, R. G. (1973), Olfactory psychophysical parameters in man, rat, dog, and pigeon. *J. comp. physiol. Psychol.*, **85**, 221–32.

Day, M. G. (1968), Food habits of British stoats (*Mustela erminea*) and weasels (*Mustela nivalis*). *J. Zool.*, **155**, 485–97.

Desjardins, C., Maruniak, J. A. and Bronson, F. H. (1973), Social rank in the house mouse: differentiation revealed by ultraviolet visualisation of urinary marking patterns. *Science*, **182**, 939–41.

Devor, M., and Murphy, M. R. (1973), The effect of peripheral olfactory blockade on the social behaviour of the male golden hamster. *Behav. Biol.*, **9**, 31–42.

Ditmars, R. L. (1910), *Reptiles of the world*, Sturgis and Walton Co., New York.

Dix, M. W. (1968), Snake food preference: Innate intraspecific geographic variation. *Science*, **159**, 1478–9.

Dixon, A. K. and Mackintosh, J. H. (1971), Effects of female urine upon the social behaviour of adult male mice. *Animal Behaviour*, **19**, 138–40.

Dixon, A. K. and Mackintosh, J. H. (1976), Olfactory mechanisms affording protection from attack to juvenile mice (*Mus musculus* L.). *Z. Tierpsychologie*, **41**, 225–34.

Dixon, A. K. and Mackintosh, J. H. (1975), The relationship between the physiological condition of female mice and the effects of their urine on the social behaviour of adult males. *Animal Behaviour*, **23**, 513–20.

Dole, J. W. (1972), The role of olfaction and audition in the orientation of leopard frogs, *Rana pipiens*. *Herpetologica*, **28**, 258–60.

van Dorp, D. A., Klok, R. and Nugteren, D. H. (1973), New macrocyclic compounds from the secretions of the civet cat and the musk rat. *Recueil*, **92**, 915–28.

Doty, R. L., Ford, M., Pretti, G. and Huggins, S. R. (1975), Changes in the intensity and pleasantness of human vaginal odors during the menstrual cycle. *Science*, **190**, 1316–18.

Doty, R. L. (1972), Odor preferences of female *Peromyscus maniculatus bairdi* for male mouse odours of *P. m. bairdi* and *P. leucopus noveboracensis* as a function of estrous state. *J. comp. physiol. Psychol.*, **81**, 191–7.

Døving, K. B., Nordeng, H. and Oakley, B. (1974), Single unit discrimination of fish odours released by char (*Salmo alpinus* L.) populations. *Comp. Biochem. Physiol.*, **47A**, 1051–63.

van Drongelen, W., Holley, A. and Døving, K. B. (1978), Convergence in the olfactory system: quantitative aspects of odour sensitivity. *J. theor. Biol.*, **71**, 39–48.

Drummond, D. C. (1970), Variation in rodent populations in response to control measures, In: *Variability in Mammalian Populations*. Symposium of the Zoological Society of London, No. 26 (ed. R. J. Berry and H. N. Southern), Academic Press, London, pp. 351–67.

Dyson, G. M. (1938), The scientific basis of odour. *Chem. Ind.*, 647–51.

Ebling, J. F. (1963), Hormonal control of sebaceous glands in experimental animals, In: *Advances in biology of skin, Vol. 4., The Sebaceous glands* (ed. W. Montagna, R. A. Ellis, A. F. Silver), Pergamon Press, New York, pp. 200–19.

Eggert, B. (1931), Die Geschlechtsorgane der Gobiiformes und Blenniformes. *Z. wissenschaft. Zool.*, **193**, 249–558.

Ehrenfeld, J. G. and Ehrenfeld, D. W. (1973), Externally secreting glands of freshwater and sea turtles. *Copeia*, 305–14.

Ehrlich, S. (1966), Ecological effects of reproduction in nutria (*Myocastor coypus* Mol.). *Mammalia*, **30**, 144–52.

Eibl-Eibesfeldt, I. (1953), Zur Ethologie des Hamsters. *Z. Tierpsychologie*, **10**, 204–54.

Eisenberg, J. F., McKay, G. M. and Jainudeen, M. R. (1971), Reproductive behaviour of the Asiatic elephant (*Elaphas maximus maximus* L.). *Behaviour*, **38**, 193–225.

Eisner, T., Connor, W. E., Hicks, K., Dodge, K. R., Rosenberg, H. T., Jones, T. H., Cohen, M. and Meinwald, J. (1977), Stink of stink-pot turtle identified: ω -Phenylalkanoic acids. *Science*, **196**, 1347–9.

Ellis, H. (1936), *Studies in the psychology of sex*, Vol. I part 3, Random House, New York.

Epple, G. (1974), Primate pheromones, In: *Pheromones* (ed. M. C. Birch), North Holland Publishing Co., Amsterdam, pp. 366–85.

Epple, G. and Lorenz, K. (1967), Vorkommen, Morphologie und Funktion der Sternaldrüse bei den Platyrrhini. *Folia Primatologica*, **7**, 98–126.

Erlinge, S. (1968), Territoriality of the otter, *Lutra lutra* L. *Oikos*, **19**, 81–98.

Evans, C. M., Mackintosh, J. H., Kennedy, J. F. and Robertson, S. M. (1978), Attempts to characterise and isolate aggression reducing olfactory signals from the urine of female mice *Mus musculus* L. *Physiol. Behav.*, **20**, 129–34.

Ewer, R. F. (1974), *The Carnivores*, Weidenfeld and Nicholson, London.

Ewer, R. F. (1968), *Ethology of Mammals*, Elek Press, London.

Ewer, R. F. (1961), Further observations on suckling behaviour in kittens, together with some general considerations of the interrelations of innate and acquired responses. *Behaviour*, **18**, 247–60.

Fletcher, I. C. and Lindsay, D. R. (1968), Sensory involvement in the mating behaviour of domestic sheep. *Animal Behaviour*, **16**, 410–4.

Fiaschi, V., Farina, A. and Ioale, P. (1974), Homing experiments on swifts *Apus apus* (L) deprived of olfactory perception. *Monitore zool. Ital. (NS)*, **8**, 235–44.

Forward, R. B., Jr., Horch, K. W. and Waterman T. H. (1972), Visual orientation at the water surface by the teleost *Zenarchopterus*. *Biol. Bull.*, **143**, 112–26.

Frädich, H. (1967), Das Verhalten der Schweine (Suidae, Tayassuidae) und Flusspferde (Hippotamidae). *Handbuch der Zoologie*, **10**, 1–44.

Friedmann, H. (1955), The honey guides. *Bull. U.S. Nat. Mus.*, **208**, 1–292.

Freud, S. (1909), *Notes upon a case of obsessional neurosis*, Standard edition, Vol. 10.

Freud, S. (1949), *Collected papers IV*, London.

Frisch, K. von (1938), Zur Psychologie das Fisch-Schwarmes. *Naturwissenschaften*, **26**, 601–6.

Gasc, J. P., Lageron, A. and Schlumberger, J. (1970), Morphologie, histologie et histochemie des glandes femorales chez un individu mâle de *Ctenosaura acantha* (Shaw) (Reptilia, Sauria, Iguanidae), suivi de réflexions sur le rôle des glandes fémorales chez les lézards. *Gegenbaurs Morphologisches Jahrbuch*, **114**, 572–90.

Gemne, G. and Døving, K. B. (1969), Ultrastructural properties of primary olfactory nerons in fish (*Lota lota* L.) *Am. J. Anat.*, **126**, 457–75.

George, C. J. (1960), Behavioral interaction of the pickerel (*Esox niger* Le Seuer and *Esox americanus* Le Seuer) and the mosquito-fish [*Gambusia partruelis* (Baird and Girard)] Ph.D. thesis, Harvard Univ., quoted in Wilson, E. O. (1968), *Chemical Systems in Animal Communication, Techniques of Study and Results of Research*, (ed. T. A. Sebeok), University of Indiana Press, Bloomington, Ind., pp. 75–102.

Getchell, M. L. and Gesteland, R. C. (1972), The chemistry of olfactory reception: Stimulus-specific protection from sulfhydryl reagent inhibition. *Proc. natn. Acad. Sci. U.S.A.*, **69**, 1494–98.

Gibson, J. J. (1966), *The senses considered as perceptive systems*, George Allen & Unwin Ltd., London.

Gilder, P. M. and Slater, P. J. B. (1978), Interest of mice in conspecific male odours is influenced by degree of kinship. *Nature*, **274**, 364–5.

Godfrey, J. (1958), The origin of sexual isolation between bank voles. *Proc. R. Phys. Soc. Edin.*, 47–55.

Gooding, R. (1963), The olfactory organ of the skipjack, *Katsuwonus pelamis*. *FAO Fish Reports*, **6**, 1621–31.

Goodrich, B. S., Hesterman, E. R., Murray, K. E., Mykytowycz, R., Stanley, G. and Sugowdz, G. (1978), Identification of behaviorally significant volatile compounds in the anal gland of the rabbit, *Oryctolagus cuniculus*. *J. Chem. Ecol.*, **4**, 581–94.

Goodrich, B. S. and Mykytowycz, R. (1972), Individual and sex differences in the chemical composition of pheromone-like substances from the skin glands of the rabbit, *Oryctolagus cuniculus*. *J. Mammal.*, **53**, 540–8.

Gorman, M. L. (1976), A mechanism for individual recognition by odour in *Herpestes auropunctatus* (Carnivora: Viverridae). *Animal Behaviour*, **24**, 141–6.

Gorman, M. L., Nedwell, D. B. and Smith, R. M. (1974), An analysis of the contents of the anal scent pockets of *Herpestes auropunctatus* (Carnivora: Viverridae). *J. Zool.*, **172**, 389–99.

Göz, H. (1941), Über den Art und Individualgeruch bei Fischen. *Z. vergleichende Physiol.*, **29**, 1–45.

Graf, R. and Meyer-Holzapfel, M. (1974), Die Wirkung von Harnmarken auf Artgenossen beim Haushund. *Z. Tierpsychologie*, **35**, 320–2.

Grant, D., Anderson, O. and Twitty, V. C. (1968), Homing orientation by olfaction in newts (*Taricha rivularis*). *Science*, **160**, 1354–56.

Grassé, P.-P. (1958), *Traité de Zoologie*, Vol. 13, Masson et Cie, Paris.

Gray, H. (1905), *Anatomy, descriptive and surgical*, 16th ed., Longmans, Green & Co., London.

Griffith, C. R. (1920), The behaviour of white rats in the presence of cats. *Psychobiology*, **2**, 19–28.

Graziadei, P. P. C. (1977), Functional anatomy of the mammalian chemoreceptor system, In: *Chemical Signals in Vertebrates*, (ed. D. Müller-Schwarze and M. M. Mozell), Plenum Publishing Corp., New York, pp. 435–54.

Graziadei, P. P. C. and Moulton, G. (1977), Continuous nerve cell renewal in the olfactory system, In: *Handbook of Sensory Physiology*, Vol. IX, (ed. M. Jacobson), Springer Verlag, New York.

Gregory, E., Engel, K. and Pfaff, D. (1975), Male hamster preference for odors of female hamster vaginal discharges: studies of experiential and hormonal determinants. *J. comp. physiol. Psychol.*, **89**, 442–6.

Grubb, T. C. (1974), Olfactory navigation to the nesting burrow in Leach's petrel (*Oceanodroma leucorrhoa*). *Animal Behaviour*, **22**, 192–202.

Grubb, J. C. (1976), Maze orientation by Mexican toads, *Bufo valliceps* (Amphibia, Anura, Bufonidae) using olfactory and configurational cues. *J. Herpetol.*, **10**, 97–105.

Grubb, J. C. (1973), Olfactory orientation in *Bufo woodhousei fowleri*, *Pseudacris clarki* and *Pseudacris streckeri*. *Animal Behaviour*, **21**, 726–32.

Grubb, T. C. (1972), Smell and foraging in shearwaters and petrels. *Nature*, **237**, 404–5.

Guggisberg, C. A. W. (1972), *Crocodiles*, Wren Publishing, Victoria, BC.

Gurney, J. H. (1922), On the sense of smell possessed by birds. *Ibis.*, **4**, 225–53.

Guthrie, R. D. and Petocz, R. G. (1970), Weapon automimicry among mammals. *Am. Nat.*, **104**, 585–8.

Hainer, R. M., Emslie, A. G. and Jacobson, A. (1954), An information theory of olfaction. *Ann. N.Y. Acad. Sci.*, **58**, 158–74.

Halpin, Z. T. (1974), Individual differences in the biological odors of the Mongolian gerbil (*Meriones unguiculatus*). *Behav. Biol.*, **11**, 253–9.

Hammer, F. J. (1951), The relation of odour, taste and flicker-fusion thresholds to food intake, *J. comp. physiol. Psychol.*, **44**, 403–11.

Hara, T. J. (1976), Structure-activity relationships of amino acids in fish olfaction. *Comp. Biochem. Physiol.*, **54**(A), 31–6.

Hara, T. J. and Macdonald, S. (1976), Olfactory responses to skin mucous substances in rainbow trout *Salmo gairdneri*. *Comp. Biochem. Physiol.*, **54**(A), 41–4.

Harden-Jones, F. R. (1968), *Fish Migration*, Edward Arnold, London.

Harrington, J. E. (1976), Discrimination between individuals by scent in *Lemur fulvus*. *Animal Behaviour*, **24**, 207–12.

Hasler, A. D. (1966), *Underwater Guideposts; Homing of Salmon*, The University of Wisconsin Press, Madison, Wis.

Hasler, A. D. (1957), The sense organs; olfactory and gustatory senses of fishes, In: *Physiology of Fishes*, Vol. II, (ed. M. E. Brown), Academic Press, London, pp. 187–209.

Hatlapa, H. H. (1977), Zur biologischen Bedeutung des Präorbitalorgans beim Rotwild, Prägung, Individualgeruch, Orientierung. *Berlin und München Tierärzliche Wochenschrift*, **90**, 100–4.

Hawes, M. L. (1976), Odor as a possible isolating mechanism in sympatric species of shrews (*Sorex vagrans* and *Sorex obscurus*). *J. Mammal.*, **57**, 404–6.

Hayashi, S., and Kimura, T. (1974), Sex-attractant emitted by female mice. *Physiol. Behav.*, **13**, 563–7.

Hediger, H. (1949), Säugetier-Territorien und ihre Markierung. *Bijdragentot de Dierkde*, **28**, 172–84.

Heimer, L. (1968), Synaptic distribution of centripetal and centrifugal nerve fibres in the olfactory system of the rat. An experimental anatomical study. *J. Anat.*, **103**, 413–32.

Hemmer, H. and Schopp, G. (1975), Significance of olfaction in the feeding behaviour of toads (Amphibia, Anura, Bufonidae): studies with the natterjack (*Bufo calamita* Laur.) *Z. Tierpsychologie*, **39**, 173–7.

Hemmings, C. C. (1966), Olfaction and vision in fish schooling. *J. exp. Biol.*, **45**, 449–64.

Hennessy, D. F. and Owings, D. H. (1978), Snake species discrimination and the role of olfactory cues in the snake-directed behaviour of the California ground squirrel. *Behaviour*, **65**, 116–24.

Henry, J. D. (1977), Use of urine marking in scavenging behaviour of red fox (*Vulpes vulpes*). *Behaviour*, **61**, 82–105.

Hersher, L., Richmond, J. B. and Moore, A. U. (1963), Modifiability of the critical period for the development of maternal behaviour in sheep and goats. *Behaviour*, **20**, 311–20.

Herter, K. (1941), Amphibia. In: *Handbuch der Zoologie*, VI, Walter de Gruyler, Berlin, pp. 200–12.

Hesterman, E. R., Goodrich, B. S. and Mykytowycz, R. (1976), Behavioural and cardiac responses of the rabbit, *Oryctolagus cuniculus*, to chemical fractions from anal glands. *J. chem. Ecol.*, **2**, 25–37.

Hesterman, E. R. and Mykytowycz, R. (1968), Some observations on the odours of anal gland secretions from the rabbit, *Oryctolagus cuniculus* (L.). *CSIRO Wildlife Res.*, **13**, 71–81.

Heusser, H. (1958), Zum geruchlichen Beutefinden und Gähnen der Kreuzkröte (*Bufo calamita* Laur.). *Z. Tierpsychologie*, **15**, 94–8.

Hillyer, G. M. (1976), An investigation using a synthetic porcine pheromone and the effect on days from weaning to conception. *Vet. Rec.*, **98**, 93–4.

Hobson, E. S. (1963), Feeding behaviour in three species of sharks. *Pacific Sci.*, **17**, 171–94.

Hodge, W. H. (1946), Camels of the clouds. *Natn. Geographic Mag.*, **89**, 641–56.

Holling, C. S. (1958), Sensory stimuli involved in the location and selection of sawfly cocoons by small mammals. *Can. J. Zool.*, **36**, 633–53.

von Holst, D. (1969), Sozialer Stress bei Tupajas (*Tupaia belangeri*). *Z. vergleichende Physiol.*, **63**, 1–58.

Holt, W. V. and Tam, W. H. (1973), Steriod metabolism by the chin gland of the male cuis, *Galea musteloides*. *J. Reproduction Fertility*, **33**, 53–9.

Howard, W. E., Marsh, R. E. and Cole, R. E. (1968), Food detection by deer mice using olfactory rather than visual cues. *Animal Behaviour*, **16**, 13–7.

Howard, W. E., Palmateer, S. D. and Marsh, R. E. (1969), A body capacitor-olfactometer for squirrels and rats. *J. Mammal.*, **50**, 771–6.

Huggins, G. R. and Preti, G. (1976), Volatile constitutents of human vaginal secretions. *Am. J. Obstet. Gyn.*, **126**, 129–36.

Hughes, G. R. (1974), The sea turtles of south-east Africa. II. The biology of the Tongaland Loggerhead turtle *Caretta caretta* L. with comments on the leatherback turtle *Dermochelys coriacea* L. and the green turtle *Chelonia mydas* L. in the study region. *S. A. Ass. Mar. Biol. Res. Oceanographic Res. Instit.* Investigational Report No. 36.

Hunter, J. R. and Hasler, A. D. (1965), Spawning association of the redfin shiner *Notropis umbratilis* and the green sunfish *Lepomis cyanellus*. *Copeia*, 265–81.

Hurley, J. K. and Shelley, W. B. (1960), *The human apocrine sweat gland in health and disease*, Thomas, Springfield, Ill.

Inkster, I. J. (1957), The mating behaviour of sheep. *N. Z. Sheepfarming Ann.*, 163–9.

Jacob, J. (1977), Composition of the ventral gland-pad sebum from the Mongolian gerbil, *Meriones unguiculatus*. *Z. Naturforsch.*, **32**, 735–8.

Jacob, J. and von Lehmann, E. (1976), Bemerkungen zu einer Nasendrüse des Sumphirsches, *Odocoileus* (*Dorcelaphus*) *dichotomus* (Illiger, 1811). *Säugetierkundliche Mitteilungen*, **24**, 151–6.

Jackson, J. E. (1974), Feeding habits of deer. *Mammal Rev.*, **4**, 93–101.

Jackson, R. T. (1960), The olfactory pigment. *J. cell. comp. Physiol.*, **55**, 143–7.

Johnson, C. E. (1921), The 'hand-stand' habit of the spotted skunk. *J. Mammal.*, **2**, 87–9.

Johnson, R. P. (1973), Scent marking in mammals. *Animal Behaviour*, **21**, 521–35.

Johnson, R. P. (1975), Scent marking with urine in two races of the bank vole (*Clethrionomys glareolus*). *Behaviour*, **55**, 81–93.

Johnston, R. E. (1974), Sexual attraction function of golden hamster vaginal secretion. *Behav. Biol.*, **12**, 111–7.

Johnston, R. E. (1975), Scent marking by male golden hamsters (*Mesocricetus auratus*). *Z. Tierpsychologie*, **37**, 75–98; 138–44; 213–21.

Johnston, R. E. (1975), Sexual excitation function of hamster vaginal secretion. *Animal Learning and Behavior*, **3**, 161–6.

Johnston, R. E. (1977), Sex pheromones in golden hamsters, In: *Chemical Signals in Vertebrates* (ed. D. Müller-Schwarze and M. M. Mozell), Plenum Publishing Corp., New York, pp. 225–49.

Jolly, A. (1966), *Lemur Behavior: a Madagascar field study*, Chicago University Press, Chicago, Ill.

Jones, R. B. and Nowell, N. W. (1973), The coagulating glands as a source of aversive and aggression-inhibiting pheromone(s) in the male albino mouse. *Physiol. Behav.*, **11**, 455–62.

Jouventin, P. (1977), Olfaction in snow petrels. *The Condor*, **79**, 498–9.

Kahmann, H. (1932), Sinnesphysiologische Studien am Reptilien – I. Experimentalle Untersuchungen über das Jacobsonische Organ der Eidechsen und Schlangen. *Zoologische Jahrbucher: Abteilung für allgemeine Zoologie und Physiologie der Tiere*, **51**, 173–238.

Kalmijn, A. J. (1971), The electric sense of sharks and rays. *J. exp. Biol.*, **55**, 371–83.

Kämper, R. and Schmidt, U. (1977), Die Morphologie der Nasenhöhle bei einigen neotropischen Chiropteren. *Zoomorphology*, **87**, 3–19.

Kalmus, H. (1955), The discrimination by the nose of the dog of individual human odours and in particular of the odours of twins. *Br. J. Anim. Behav.*, **111**, 25–31.

Kalogerakis, M. G. (1963), The role of the olfaction in sexual development. *Psychosomatic Med.*, **25**, 420–32.

Kaufman, G. W., Siniff, D. B. and Reichle, R. (1975), Colony behavior of Weddell seals, *Leptonychotes weddelli*, at Hutton Cliffs, Antarctica. *Rapports et procès-verbaux des réunions de la conseil permanent international pour l'exploration de la mer*, **169**, 228–46.

Keenleyside, M. H. A. (1955), Some aspects of the schooling behaviour of fish. *Behaviour*, **8**, 183–248.

Keith, L., Stromberg, P., Krotoszynski, B. K., Shah, J. and Dravnieks, A. (1975), The odors of the human vagina. *Archiv für Gynäkologie*, **220**, 1–10.

Kennelly, J. J., Johns, B. E., and Garrison, M. V. (1972), Influence of sterile males on female fecundity of a rat colony. *J. Wildlife Management*, **36**, 161–5.

Kiley-Worthington, M. (1965), The Waterbuck (*Kobus defassa* Ruppel 1835 and *K. ellipsiprimnus* Ogilby 1833) in East Africa: spatial distribution. A study of the sexual behaviour. *Mammalia*, **29**, 177–204.

Kleerekoper, H. (1969), *Olfaction in Fishes*, Indiana University Press, Bloomington, Indiana.

Kleerekoper, H. and Morgensen, J. (1963), Role of olfaction in the orientation of *Petromyzon marinus*. I. Response to a single amine in the prey's body odor. *Physiol. Zool.*, **36**, 347–60.

Klopfer, P. H., Adams, D. K. and Klopfer, M. R. (1962), Maternal "imprinting" in goats. *Proc. natn. Acad. Sci. U.S.A.*, **52**, 911–14.

Knappe, H. (1964), Zur Funktion des Jacobsonschen Organs (Organon vomeronasale Jacobsoni). *Zoologischer Garten*, **28**, 188–94.

Koch, A. L., Carr, A. and Ehrenfeld, D. W. (1969), The problem of open-sea navigation: the migration of the green turtle to Ascension Island. *J. theor. Biol.*, **22**, 163–79.

Kolb, A. (1958), Uber die Nahrungsaufnahme einheimischer Fledermäuse von Boden. *Verhandlungen der Deutschen zoologischen Gesellschaft*, **22**, 162–8.

Kolb, A. (1961), Sinnesleistungen einheimischer Fledermäuse bei der Nahrungssuche und Nahrungsauswahl auf dem Boden und in der Luft. *Z. vergleichende Physiol.*, **44**, 550–64.

Kovach, J. K. and Kling, A. (1967), Mechanisms of neonate sucking behaviour in the kitten. *Animal Behaviour*, **15**, 91–101.

Krames, L. (1970), Responses of female rats to the individual body odours of male rats. *Psychonomic Sci.*, **20**, 274–5.

Krames, L., Carr, W. and Bergman, B. (1969), A pheromone associated with social dominance among male rats. *Psychonomic Sci.*, **16**, 11–2.

Krebs, J. R. and Davies, N. B. (1978), *Behavioural Ecology – An Evolutionary Approach*, Blackwell Scientific Publications, Oxford.

Kruska, D. and Stephan, H. (1973), Volumenvergleich allokortikaler Hirnzentren bei Wild-und Hausschwein. *Acta anatomica*, **84**, 387–415.

Kruuk, H. (1972), *The spotted hyaena*, Chicago University Press, Chicago, Ill.

Kruuk, H. (1978), Spatial organization and territorial behaviour of the European badger *Meles meles*. *J. Zool.*, **184**, 1–19.

Kruuk, H. and Hewson, R. (1978), Spacing and foraging of otters (*Lutra lutra*) in a marine habitat. *J. Zool.*, **185**, 205–12.

Kubie, J. and Halpern, M. (1975), Laboratory observations of trailing behaviour in garter snakes. *J. comp. physiol. Psychol.*, **89**, 667–74.

Kühme, W. (1961), Beobachtungen am afrikanischen Elephanten (*Loxodonta africana* Blumenbach 1797) in Gefangenschaft. *Z. Tierpsychologie*, **18**, 285–96.

Kühme, W. (1963), Chemisch ausgelöste Brutpflege und Schwarmreaktionen bei *Hemichromis bimaculatus*. *Z. Tierpsychologie*, **20**, 688–704.

Kühme, W. (1964), Eine chemisch ausgelöste Schwarmreaktion bei jungen Cichliden (Pisces). *Naturwissenschaften*, **51**, 120–1.

Kulzer, E. (1961), Uber die Biologie der Nil-Flughund (*Rousettus aegyptiacus*). *Natur. Volk* (Frankfurt), **91**, 219–28.

Larsson, K. (1971), Impaired mating performances in male rats after anosmia induced peripherally or centrally. *Brain, Behavior and Evolution*, **4**, 463–71.

Lederer, E. (1950), Odeurs et parfums des animaux. *Fortschritte der Chemie organischer Naturstoffe*, **6**, 87–153.

van der Lee, S. and Boot, L. M. (1955), Spontaneous pseudopregnancy in mice. *Acta physiologica et pharmacologica Neerlandica*, **4**, 442–4.

Lehner, P. N., Krumm, R. and Cringan, A. T. (1976), Tests for olfactory repellents for coyotes and dogs. *J. Wildlife Management*, **40**, 145–50.

Le Magnen, J. (1959), Le rôle des stimuli olfacto-gustatifs dans la régulation du comportement alimentaire du mammifère. *J. Psychol. normal. path.*, **56**, 137–60.

Leon, M. (1975), Dietary control of maternal pheromone in the lactating rat. *Physiol. Behav.*, **14**, 311–9.

Leon, M. and Behse, J. H. (1977), Dissolution of the pheromonal bond: waning of approach response by weanling rats. *Physiol. Behav.*, **18**, 393–8.

Leon, M., Galef, B. G. and Behse, J. H. (1977), Establishment of pheromonal bonds and diet choice in young rats by odor pre-exposure. *Physiol. Behav.*, **18**, 387–92.

Linhart, S. B. (1973), *Predator Survey. Western US*. US Fish and Wildlife Service, Wildlife Research Center, Denver, Co.

Lindsay, D. R. (1965), The importance of olfactory stimuli in the mating behaviour of the ram. *Animal Behaviour*, **13**, 75–8.

Lindsay, D. R. (1966), Mating behaviour of ewes and its effect on mating efficiency. *Animal Behaviour*, **14**, 419–24.

Lindsay, D. R. and Robinson, T. J. (1961), Studies on the efficiency of mating in the sheep. *J. agric. Sci.*, **57**, 137–40 and 141–5.

Lisk, R. D., Zeiss, J. and Ciaccio, L. A. (1972), The influence of olfaction on sexual behavior in the male golden hamster (*Mesocricetus auratus*). *J. exp. Zool.*, **181**, 69–78.

Lockie, J. D. (1952), A comparison of some aspects of the retinae of the Manx shearwater, fulmar, petrel and house sparrow. *Q. J. micros. Sci.*, **93**, 347–56.

Lombardi, J. R. and Vandenbergh, J. G. (1977), Pheromonally induced sexual maturation in females: regulation by the social environment of the male. *Science*, **196**, 545–6.

Longhurst, W. M., Oh, H. K., Jones, M. B. and Kepner, R. E. (1968), A basis for the palatability of deer forage plants. *Trans. N. Am. Wildlife Conf.*, **33**, 181–92.

Loop, M. S. and Scoville, S. A. (1972), Response of newborn *Eumeces inexpectatus* to prey-object extracts. *Herpetologica*, **28**, 254–6.

Lorenz, K. (1943), Die angeborenen Formen möglicher Erfahrung. *Z. Tierpsychologie*, **5**, 235–409.

Lukas, T., Berner, E. S. and Kanakis, C. (1977), Diagnosis by smell? *J. med. Ed.*, **52**, 349–50.

Lydell, K. and Doty, R. L. (1972), Male rat of odor preferences for female urine as a function of sexual experience, urine age, and urine source. *Hormones and Behavior*, **3**, 205–12.

McCann, S. McD., Dhariwal, A. P. S. and Porter, J. C. (1968), Regulation of the adenohypophysis. *A. Rev. Physiol.*, **30**, 589–640.

McCarley, H. (1964), Ethological isolation in the cenospecies *Peromyscus leucopus*. *Evolution*, **18**, 331–2.

McCarty, R. and Southwick, C. H. (1977), Cross-species fostering: Effects on the olfactory preference of *Onychomys torridus* and *Peromyscus leucopus*. *Behav. Biol.*, **19**, 255–60.

McCleave, J., Rommel, S. and Cathcart, C. (1971), Weak electric and magnetic fields in fish orientation, In: *Orientation: Sensory Basis*, (ed. H. Adler), *Ann. N. Y. Acad. Sci.*, **188**, 270–82

McClintock, M. K. (1971), Menstrual synchtomy and suppression. *Nature*, **229**, 244–5.

McCrady, E. (1938), The embryology of the opossum. *Memoirs of the Wistar Institute of Anatomy and Biology*, **16**, 1–234.

Macdonald, D. W. (in press), Patterns of scent marking with urine and faeces amongst carnivore communities, In: *Olfaction in mammals*, (ed. D. M. Stoddart), *Symp. Zool. Soc. Lond.*, Academic Press, London.

Macdonald, J. D. (1960), Secondary external nares of the gannet. *Proc. Zool. Soc. Lond.*, **135**, 357–63.

McIntosh, T. K. and Drickamer, L. C. (1977), Excreted urine, bladder urine, and the delay of sexual maturation in female house mice. *Animal Behaviour*, **25**, 999–1004.

Maderson, P. F. A. (1970), Lizard glands and lizard hands: models for evolutionary study. *Forma et Functio*, **3**, 179–203.

Madison, D. M. (1975), Intraspecific odor preferences between salamanders of the same sex: dependence on season and proximity of residence. *Can. J. Zool.*, **53**, 1356–61.

Madison, D. M. (1969), Homing behaviour of the red-cheeked salamander, *Plethodon jordani*. *Animal Behaviour*, **17**, 25–39.

Madison, D. M. (1977), Chemical communication in amphibians and reptiles, In: *Chemical Signals in Vertebrates*, (ed. D. Müller-Schwarze M. M. Mozell), Plenum Publishing Corp., New York, pp. 135–68.

Madison, D. M., Schotz, A. T., Cooper, J. C., Horrall, R. M., Hasler, A. D. and Dizon, A. E. (1973), 1. Olfactory hypotheses and salmon migration: a synopsis of recent findings. Technical Report No. 414, Fisheries Research Board of Canada.

Mahmoud, I. Y. (1967), Courtship behavior and sexual maturity in four species of kinosternid turtles. *Copeia*, 314–9.

Mann, G. (1961), Bulbus olfactorius accessorius in chiroptera. *J. comp. Neurol.*, **116**, 135–41.

Mansfield, A. W. (1958), The breeding behaviour and reproductive cycle of the Weddell seal (*Leptonychotes weddelli* Lesson); Falkland Islands Dependences Survey Scientific Reports No. 18, 1–41.

Marr, J. N. and Gardner, L. E. (1965), Early olfactory experience and later social behavior in the rat: preference, sexual responsiveness and care of young. *J. gen. Psychol.*, **107**, 167–74.

Marshall, N. B. (1967), The olfactory organs of bathypelagic fishes. In: Symposium Zoological Society No. 19, (ed. N. B. Marshall), Academic Press, London, pp. 57–70.

Martin, R. D. (1968), Reproduction and ontogeny in tree shrews (*Tupaia belangeri*) with reference to their general behaviour and taxonomic relationships. *Z. Tierpsychologie*, **25**, 409–95 and 505–32.

Mayr, E. (1963), *Animal species and evolution*, Harvard University Press, Cambridge, Mass.

Mech, D. L. (1970), *The Wolf: the ecology and behavior of an endangered species*, Natural History Press, Garden City, New York.

Menco, B. P. M. (1977), A qualitative and quantitative investigation of olfactory and nasal respiratory mucosal surfaces of cow and sheep based on various ultrastructural and biochemical methods. *Meded. Landbouwhogeschool Wageningen*, **77-13**, 1–157.

Mertl, A. S. (1975), Discrimination of individuals by scent in a primate. *Behav. Biol.*, **14**, 505–9.

Mertl, A. S. (1976), Olfactory and visual cues in social interactions of *Lemur catta*. *Folia primatologica*, **26**, 151–61.

Mertl, A. S. (1977), Habituation to territorial scent marks in the field by *Lemur catta*. *Behav. Biol.*, **21**, 500–7.

Michael, R. P. and Bonsall, R. W. (1977), Chemical signals and primate behavior, In: *Chemical Signals in Vertebrates* (ed. D. Müller-Schwarze and M. M. Mozell), Plenum Publishing Corp., New York, pp. 251–72.

Michael, R. P., Bonsall, R. W. and Kutner, M. (1975), Volatile fatty acids, "copulins" in human vaginal secretions. *Psychoneuroendocrinology*, **1**, 153–63.

Michael, R. P., Bonsall, R. W. and Warner, P. (1974), Human vaginal secretions: volatile fatty acid content. *Science*, **186**, 1217–9.

Michael, R. P., Zumpfe, D., Keverne, E. B. and Bonsall, R. W. (1972), Neuroendocrine factors in the control of primate behavior. *Rec. Prog. Hormone Res.*, **78**, 665–705.

Miles, S. G. (1968), Rheotaxis of elvers of the American eel (*Auguilla rostrata*) in the laboratory to

water from different streams in Nova Scotia. *J. Fisheries Res. Board Can.*, **25**, 1591–1602.
Millais, J. E. (1895), *A breath from the Veldt*, Henry Sotheran & Co, London.
Moltz, H. and Leidahl, L. C. (1977), Bile, prolactin and maternal pheromone. *Science*, **196**, 81–3.
Moltz, H. and Leon, M. (1973), Stimulus control of the maternal pheromone in the lactating rat. *Physiology and Behavior*, **10**, 69–71.
Mollenauer, S., Plotnik, R. and Snyder, E. (1974), Effects of olfactory bulb removal on fear responses and passive avoidance in the rat. *Physiology and Behavior*, **12**, 141–4.
Moller, W. (1932), Das Epithel der Speiseröhrenschleimhaut der blütenbescichender Fledermaus *Glossophaga soricina* im Vergleich zu insekten fressenden Chiroptera. *Z. mikros.-anat. Forsch.*, **29**, 637–53.
Moore, R. E. (1965), Olfactory discrimination as an isolating mechanism between *Peromyscus maniculatus* and *Peromyscus polionotus*. *Am. Midland Nat.*, **73**, 85–100.
Moulton, D. G. (1960), Studies in olfactory acuity 5. The comparative olfactory sensitivity of pigmented and albino rats. *Animal Behaviour*, **8**, 129–33.
Moulton, D. G. (1974), Dynamics of cell populations in the olfactory epithelium. *Ann. N. Y. Acad. Sci.*, **237**, 52–61.
Moulton, D. G. and Tucker, D. (1964), Electrophysiology of the olfactory system. *Ann. N. Y. Acad. Sci.*, **116**, 380–428.
Muller, K. and Kiepenheuer, J. (1976), Olfactory orientation in frogs. *Naturwissenschaften*, **63**, 49–50.
Müller-Schwarze, D. (1971) Pheromones in black-tailed deer (*Odocoileus hemionus columbianus*). *Animal Behaviour*, **19**, 141–52.
Müller–Schwarze, D., Müller–Schwarze, C., Singer, A. C. and Silverstein, R. M. (1974), Mammalian pheromone: identification of active component in the subauricular scent of the male pronghorn. *Science*, **183**, 860–2.
Müller–Schwarze, D., Quay, W. B. and Brundin, A. (1977), The caudal gland in reindeer (*Rangifer tarandus* L): its behavioural role, histology and chemistry. *J. chem. Ecol.*, **3**, 591–602.
Müller–Schwarze, D., Volkman, N. J. and Zemanek, K. F. (1977), Osmetrichia: specialised scent hair in black tailed deer. *J. Ultrastruct. Res.*, **59**, 223–30.
Müller–Velten, H. (1966), Uber den Angstgeruch bei der Hausmaus (*Mus musculus* L.). *Z. vergleichende Physiologie*, **52**, 401–29.
Mulligan, J. A. (1966), Singing behavior and its development in the song sparrow *Melospiza melodia*. *Univ. Cal. Publ. Zool.*, **81**, 1–76.
Muul, I. (1970), Day length and food caches, In: *Field Studies in Natural History*, Van Nostrand Reinhold, New York.
Mykytowycz, R. (1965), Further observations on the territorial function and histology of the submandibular cutaneous (chin) glands in the rabbit, *Oryctolagus cuniculus* (L.). *Animal Behaviour*, **13**, 400–12.
Mykytowycz, R. (1968), Territorial marking by rabbits. *Sci. Am.*, **218**, 116–26.
Mykytowycz, R. and Dudzinski, M. L. (1966), A study of the weight of odoriferous and other glands in relation to social status and degree of sexual activity in the wild rabbit, *Oryctolagus cuniculus* (L.). *CSIRO Wildlife Res.*, **11**, 31–47.
Mykytowycz, R. and Dudzinski, M. L. (1972), Aggressive and protective behavior of adult rabbits, *Oryctolagus cuniculus* (L) towards juveniles. *Behaviour*, **43**, 7–120.
Mykytowycz, R. and Gambale, S. (1969), The distribution of dung-hills and the behaviour of free-living rabbits, *Oryctolagus cuniculus* (L) on them. *Forma et Functio*, **1**, 333–49.
Mykytowycz, R. and Hesterman, E. R. (1970), The behaviour of captive wild rabbits, *Oryctolagus cuniculus* (L) in response to strange dung-hills. *Forma et Functio*, **2**, 1–12.
Mykytowycz, R., Hesterman, E. R., Gambale, S. and Dudzinski, M. L. (1976), A comparison of the effectiveness of the odors of rabbits, *Oryctolagus cuniculus*, in enhancing territorial confidence. *J. chem. Ecol.*, **2**, 13–24.
Myrberg, A. A. (1966), Parental recognition of young cichlid fishes. *Animal Behaviour*, **14**, 565–71.

Nelson, J. E. (1965), Behaviour of Australian Pteropodidae (Megachiroptera). *Animal Behaviour*, **13**, 544–57.

Negus, V. E. (1958), *Comparative Anatomy and Physiology of the Nose and Paranasal Sinuses*, E. & S. Livingstone, London.

Nevo, E., Bodmer, M. and Heth, G. (1976), Olfactory discrimination as an isolating mechanism in speciating mole rats. *Experientia*, **32**, 1511–2.

Nevo, E., Naftali, G. and Guttman, R. (1975), Aggression patterns and speciation. *Proc. natn. Acad. Sci. U.S.A.*, **72**, 3250–4.

Newcombe, C. and Hartman, G. (1973), Some chemical signals in the spawning behaviour of the rainbow trout (*Salmo gairdneri*). *J. Fisheries Res. Board Can.*, **30**, 995–7.

Nicolaides, N. (1974), Skin lipids; their biochemical uniqueness. *Science*, **186**, 19–26.

Nieuwenhuys, R. (1967), Comparative anatomy of olfactory centres and tracts, In: *Progress in Brain Research*, Vol 23 (ed. Y. Zotterman), Elsevier Amsterdam.

Noble, G. K. (1931), *Biology of the Amphibia*, McGraw-Hill, New York.

Noble, G. K. (1937), The sense organs involved in the courtship of *Storeria*, *Thamnophis* and other snakes. *Bull. Am. Mus. Nat. History*, **73**, 673–725.

Noble, G. K. and Clausen, H. J. (1936), The aggregation behavior of *Storeria dekayi* and other snakes with especial reference to the sense organs involved. *Ecol. Monographs*, **6**, 269–316.

Nordeng, H. (1971), Is the local orientation of anadromous fishes determined by pheromones? *Nature*, **233**, 411–3.

Nováková, V. and Dlouhá, H. (1960), Effect of severing the olfactory bulbs on the intake and excretion of water in the rat. *Nature*, **186**, 638–9.

Obst, C. and Schmidt, U. (1976), Untersuchungen zum Riechvermögen von *Myotis myotis* (Chiroptera). *Z. Säugetierkunde*, **41**, 101–8.

Oldak, P. D. (1976), Comparison of the scent gland secretion lipids of twenty five snakes: Implications for biochemical systematics. *Copeia*, 320–6.

Ottoson, D. (1958), The slow electrical response of the olfactory end organs. *Exp. Cell Res.*, **5**, 451–69.

Ottoson, D. (1958), Studies on the relationship between olfactory stimulating effectiveness and physico-chemical properties of odorous compounds. *Acta physiologica scandinavica*, **43**, 167–81.

Ottoson, S. (1963), Some aspects of the function of the olfactory system, *Pharmacol. Rev.*, **15**, 1–42.

Pagies, E. (1968), Glandes odorantes des pangolins. *Biol. Gabon.*, **4**, 353–400.

Papi, F. (1976), The olfactory navigation system of the homing pigeon. *Verhandlungen der Deutschen zoologischen Gesellschaft*, 1976, 184–205.

Paris, P. (1914), Recherches sur la glande uropygienne des oiseaux. *Arch. zoologie expérimentale et générale*, **53**, 139–276.

Parsons, S. D. and Hunter, G. L. (1967), Effect of the ram on duration of oestrus in the ewe. *J. Reproduction and Fertility*, **14**, 61–70.

Partridge, B. L., Liley, N. R. and Stacey, N. E. (1976), The role of pheromones in the sexual behaviour of the goldfish. *Animal Behaviour*, **24**, 291–300.

Passy, M. J. (1892), L'odeur dans la série des alcools. *Comptes rendus des séances de la Société de biologie*, **44**, 447–55.

Patterson, R. L. S. (1968), Identification of 3α-hydroxy-5α-androst-16-ene as the musk odour component of boar submaxillary salivary gland and its relationship to the sex odour taint in pork meat. *J. Sci. Food Agric.*, **19**, 434–8.

Perry, G. C., Patterson, R. L. S. and Stinson, G. E. (1973), Submaxillary salivary gland involvement in porcine mating behaviour, In: *Proc. VIIth International Cong. Animal Reproduction and A. I.*, Munich.

Peters, R. P. and Mech, L. D. (1975), Scent marking in wolves. *Am. Sci.*, **63**, 628–37.

Petter, J. J. (1965), The Lemurs of Madagascar, In: *Primate Behaviour* (ed. I. de Vore), Holt, Rinehart and Winston, New York.

Pfaff, D. and Gregory, E. (1971), Olfactory coding in the olfactory bulb and medial forebrain bundle of normal and castrated male rats. *J. Neurophysiol.*, **34**, 208–16.

Pfaff, D. and Pfaffman, C. (1969), Behavioral and electrophysiological responses of male rats to female rat urine odors, In: *Olfaction and Taste III* (ed. C. Pfaffman), Rockefeller University Press, New York, pp. 258–67.

Pfaffman, C. (1971), Sensory reception of olfactory cues. *Biol. Reprod.*, **4**, 327–43.

Pfeiffer, W. (1963), Alarm substances. *Experientia*, **19**, 113–23.

Pfeiffer, W. (1966), Die Verbreitung der Schreckreaktion bei Kaulquappen und die Herkunft des Schreckstoffes. *Z. vergleichende Physiologie*, **52**, 79–98.

Pfeiffer, W. (1974), Pheromones in fish and amphibia, In: *Pheromones*, (ed. M. C. Birch), North-Holland Publishing Co., Amsterdam, pp. 269–96.

Pfeiffer, W. and Lemke, J. (1973), Untersuchungen zur Isolierung und Identifizierung des Schreckstoffes aus der Haut der Elritze, *Phoxinus phoxinus* (L) (Cyprinidae, Ostariophysi, Pisces). *J. comp. Physiol.*, **82**, 407–10.

Phillipson, J. (1967), *Ecological Energetics*, Edward Arnold, London.

Pilters, H. (1956), Das Verhalten der Tylopoden, *Handbuch der Zoologie VIII*, **19**, 1–24.

Planel, H. (1954), Etudes sur la physiologie de l'organe de Jacobson. *Arch. d'anatomie, d'histologie et d'embryologie*, **35**, 199–205.

Poduschka, W. and Firbas, W. (1969), Das Selbsbespeicheln des Igel *Erinaceus europaeus roumanicus* Linn. 1758 steht in Beziehung zur Funktion des Jacobsonschen Organes. *Z. Säugetierkunde*, **33**, 160–70.

Porsch, O. (1932), Das Problem Fledermausblume. *Auzeiger der Akademie Wissenschaften*, Wien No. 3.

Porter, R. H., Deni, R. and Doane, H. M. (1977), Responses of *Acomys cahirinus* pups to chemical cues produced by a foster species. *Behav. Biol.*, **20**, 244–51.

Porter, R. H. and Doane, H. M. (1976), Maternal pheromone in the spiny mouse (*Acomys cahirinus*). *Physiol. Behav.*, **16**, 75–8.

Porter, R. H. and Etscorn, F. (1974), Olfactory imprinting resulting from brief exposure in *Acomys cahirinus*. *Nature*, **250**, 732–3.

Portmann, A. (1961), Sensory organs: Part I. Skin, taste and olfaction. In: *Biology and Comparative Physiology of Birds*, 1 (ed. A. J. Marshall), Academic Press, London, pp. 37–48.

Powers, J. B. and Winans, S. S. (1975), Vomeronasal organ: critical role in mediating sexual behavior of the male hamster. *Science*, **187**, 961–3.

Preti, G., Muetterties, E. L., Furman, J., Kenelly, J. J. and Johns, B. E. (1976), Volatile constituents of dog (*Canis familiaris*) and coyote (*Canis latrans*) anal sacs. *J. chem. Ecol.*, **2**, 177–87.

Purvis, K. and Haynes, N. B. (1978), Effect of the odour of female rat urine on plasma testosterone concentrations in male rats. *J. Reproduction and Fertility*, **53**, 63–5.

Quay, W. B. (1953), *Dipodomys merriami*: Dorsal gland activity and size. *J. Mammal.*, **34**, 1–14.

Quay, W. B. (1953), Seasonal and sexual differences in the dorsal skin gland of the kangaroo rat (*Dipodomys*). *J. Mammal.*, **34**, 1–14.

Ralls, K. (1971), Mammalian scent marking. *Science*, **171**, 443–9.

Rasa, O. A. (1973), Marking behaviour and its social significance in the African dwarf mongoose, *Helogale undulata rufula*. *Z. Tierpsychologie*, **32**, 293–318.

Rechav, Y., Norval, R. A. I., Tannock, J. and Colborne, J. (1978), Attraction of the tick *Ixodes neitzi* to twigs marked by the klipspringer antelope. *Nature*, **275**, 310–1.

Reed, J. R., Wieland, W. and Kimbrough, T. D. (1972), A study on the biochemistry of alarm substances in fish. *Proc. 26th Ann. Conf. S. E. Ass. Game and Fish Commissioners*, Knoxville, Tenn., pp. 608–10.

Regnier, F. E. and Goodwin, M. (1977), On the chemical and environmental modulation of pheromone release from vertebrate scent marks, In: *Chemical Signals in Vertebrates*. (ed. D. Müller-Schwarze and M. M. Mozell), Plenum Publishing Corp., New York, pp. 115–34.

Romanes, J. G. (1885), *Mental Evolution in Animals*, Routledge and Kegan Paul, London.

Rommel, S. A. Jr., and McCleave, J. D. (1972), Oceanic electric fields: Perception by American eels? *Science*, **176**, 1233–5.

Ropartz, P. (1966), Mise en évidence du rôle d'une sécrétion odorante des glandes sudoripares dans la régulation de l'activité locomotrice chez la souris. *Compte rendu hebdomadaire des séances de l'Académie des sciences, Paris*, **263**, 525–8.

Ropartz, P. (1968), The relation between olfactory stimulation and aggressive behaviour in mice. *Animal Behaviour*, **16**, 97–100.

Rose, F. L. (1970), Tortoise chin gland fatty acid composition; behavioral significance. *Comp. Biochem. Physiol.*, **22A**, 577–80.

Rottman, S. J. and Snowdon, C. T. (1972), Demonstration and analysis of an alarm pheromone in mice. *J. Comp. Physiol. Psychol.*, **81**, 483–90.

Rougier, Y. (1973), Capacité sensorielle et niveau d'activité sexuelle chez le bélier Préalpes et le beliér Ile-de-France. *Compte rendu hebdomadaire des séance de l'Académie des sciences, Paris*, **276**D, 3203–6.

Sansone, G. and Hamilton, J. G. (1969), Glyceryl ether, wax ester and triglyceride composition of the mouse preputial gland. *Lipids*, **4**, 435–40.

Satli, M. A. and Aron, C. (1976), New data on olfactory control of estral receptivity of female rats. *Compte rendu hebdomadaire des séances de l'Académie des sciences, Paris*, **282**, 875–7.

Schaffer, J. (1940), *Die Hautdrüsenorgane der Säugetiere*, Urban und Schwarzenberg, Berlin.

Schapiro, S. and Salas, M. (1970), Behavioral response of infant rats to maternal odor. *Physiol. Behav.*, **5**, 815–7.

Schildknecht, H., Witz, I., Enzmann, F., Grund, N. and Ziegler, M. (1976), Mustelan, the malodorous substance from the anal gland of the mink (*Mustela vison*) and the polecat (*Mustela putorius*). *Angewandte Chemie*, **15**, 242–3.

Schmidt, U. (1975), Vergleichende Riechschwellenbestimmungen bei neotropischen Chiropteren (*Desmodus rotundus*, *Artibeus lituratus*, *Phyllostomus discolor*). *Z. Säugetierkunde*, **40**, 269–98.

Schmidt, U. and Greenhall, A. M. (1971), Untersuchungen zur geruchlichen Orientierung der Vampirfledermäuse (*Desmodus rotundus*). *Z. vergleichende Physiologie*, **74**, 217–26.

Schultze-Westrum, T. G. (1965), Innerartliche Verständigung durch Düfte beim Gleitbeutler *Petaurus breviceps papuanus* Thomas (Marsupialia, Phalangeridae). *Z. vergleichende Physiologie*, **50**, 151–220.

Scott, J. W. and Pfaffman, C. (1967), Olfactory input to the hypothalamus: electrophysiological evidence. *Science*, **158**, 1592–4.

Seifert, K. (1970), The ultrastructure of the olfactory epithelium in macrosmatics. An electron microscopical investigation. *Normale und Pathologische Anatomie Heft*, **21**, (ed. W. Bargmann and W. Doerr), Georg Thieme Verlag, Stuttgart.

Seton, E. T. (1927), *Lives of Game Animals*, Vol III, Doubleday, New York.

Sewell, G. D. (1967), Ultrasound in rodents. *Nature*, **217**, 682–3.

Shaller, G. (1967), *The Deer and the Tiger*, Chicago University Press, Chicago, Ill.

Shank, C. C. (1972), Some aspects of social behaviour in a population of feral goats (*Capra hircus* L.) *Z. Tierpsychologie*, **30**, 488–528.

Shelton, M. (1960), Influence of the presence of a male goat on the initiation of oestrous cycling and ovulation of Angora does. *J. Animal Sci.*, **19**, 368–75.

Shumake, S. A. (1977), The search for applications of chemical signals in wildlife management, In: *Chemical Signals in Vertebrates* (ed. D. Müller-Schwarze and M. M. Mozell), Plenum Publishing Corp., New York, pp. 359–76.

Signoret, J. P. (1976), Chemical communication and reproduction in domestic mammals, In: *Mammalian Olfaction, Reproductive Processes and Behavior.* (ed. R. L. Doty), Academic Press, New York, pp. 243–56.

Signoret, J. P. and Bariteau, J. (1975), Utilisation de différents produits odorants de synthèse pour faciliter la détection des chaleurs chez la truie. *Ann. Zootechnie*, **24**, 639–43.

Singer, A. G., Agosta, W. C., O'Connell, R. J., Pfaffman, C., Bowen, D. V. and Field, F. H. (1976), Dimethyl disulfide: an attractant pheromone in hamster vaginal secretion. *Science*, **191**, 948–50.

Skeen, J. T. and Thiessen, D. D. (1977), Scent of gerbil cuisine. *Physiol. Behav.*, **19**, 11–4.

Smith, M. A. (1938), The nucho-dorsal gland of snakes. *Proc. Zool. Soc. Lond.*, **108**, 575–80.

Smith, R. J. F. (1977), Chemical communication as adaptation: alarm substance of fish, In: *Chemical Signals in Vertebrates* (ed. D. Müller-Schwarze and M. M. Mozell), Plenum Publishing Corp., New York, pp. 303–20.

Smith, W. J. (1977), *The Behavior of Communicating*, Harvard University Press, Cambridge, Mass.

Smyth, N. (1970), On the existence of 'pursuit invitation' signals in mammals. *Am. Nat.*, **104**, 491–4.

Snow, D. W. (1961), The natural history of the oilbird, *Steatornis caripensis*, in Trinidad, W I. *Zoologica*, **46**, 27–48.

Soane, I. D. and Clarke, B. (1973), Evidence for apostatic selection by predators using olfactory cues. *Nature*, **241**, 62–4.

Solomon, D. J. (1973), Evidence for pheromone-influenced homing by migrating Atlantic salmon, *Salmo salar* (L). *Nature*, **244**, 231–2.

Sokolov, V. E., Brundin, A. and Zinkevich, E. P. (1977), Differences in the chemical composition of skin gland secretion in reindeer (*Rangifer tarandus*). *Doklady Akademii nauk USSR*, **237**, 1529–32.

Sokolov, V. E. and Khorlina, I. M. (1976), Mammalian pheromones: study of the volatile acid composition of vaginal secretions in the mink *Mustela vison* Briss. *Doklady Akademii nauk USSR*, **228**, 225–7.

Sokolov, V. E., Khorlina, I. M., Glovnya, R. V. and Zhuravleva, I. L. (1974), Change in the composition of amines in the volatile substances of the vaginal secretions of American mink (*Mustela vison*) depending on the sexual cycle. *Doklady Akademii nauk USSR*, **216**, 220–2.

Solomon, D. J. (1977), A review of chemical communication in freshwater fish. *J. Fisheries Biol.*, **11**, 363–76.

Somers, P. (1973), Dialects in southern Rocky Mountain pikas, *Ochotona princeps* (Lagomorpha). *Animal Behaviour*, **21**, 124–37.

Sowls, L. K. (1974), Social behaviour of the collared peccary *Dicotyles tajacu* (L), In: *The Behaviour of Ungulates and its Relation to Management* (ed. V. Geist and F. Walther), IUCN Morges, pp. 144–65.

Stager, K. E. (1967), Avian olfaction. *Am. Zool.*, **7**, 415–9.

Stager, K. E. (1964), The role of olfaction in food location by the turkey vulture (*Cathartes aura*). *LA County Museum Contributions in Science*, No. 81.

Stehn, R. A. and Richmond, M. E. (1975), Male-induced pregnancy termination in the prairie vole, *Microtus ochrogaster*. *Science*, **187**, 1211–3.

Steiner, A. L. (1973), Self- and allo-grooming behavior in some ground squirrels (Sciuridae); a descriptive study. *Can. J. Zool.*, **51**, 151–61.

Stephan, H. (1965), Der Bulbus olfactorius accessorius bei Insektivoren und Primaten. *Acta anatomica*, **62**, 215–53.

Sternthal, D. E. (1974), Olfactory and visual cues in the feeding behavior of the leopard frog (*Rana pipiens*). *Z. Tierpsychologie*, **34**, 239–46.

Steven, D. M. (1959), Studies on the shoaling behaviour of fish. I Responses of two species to changes in illumination and to olfactory stimuli. *J. exp. Biol.*, **36**, 261–80.

Stoddart, D. M. (1970), Individual range, dispersion and dispersal in a population of water voles (*Arvicola terrestris* (L)). *J. Animal Ecol.*, **39**, 403–25.

Stoddart, D. M. (1971), Breeding and survival in a population of water voles. *J. Animal Ecol.*, **40**, 487–94.

Stoddart, D. M. (1973), Preliminary characterisation of the caudal organ secretion of *Apodemus flavicollis*. *Nature*, **246**, 501–3.

Stoddart, D. M. (1976a), *Mammalian Odours and Pheromones*, Edward Arnold, London.

Stoddart, D. M. (1976b), Effect of the odour of weasels (*Mustela nivalis* L) on trapped samples of their prey. *Oecologia*, **22**, 439–41.

Stoddart, D. M. (1977), Two hypotheses supporting the social function of odorous secretions of some Old World rodents, In: *Chemical Signals in Vertebrates* (ed. D. Müller-Schwarze and M. M. Mozell), Plenum Publishing Corp., New York, pp. 333–56.

Stoddart, D. M., Aplin, R. T. and Wood, M. J. (1975), Evidence for social difference in the flank organ secretion of *Arvicola terrestris* (Rodentia: Microtinae). *J. Zool.*, **177**, 529–40.

Strauss, E. L. (1970), A study on olfactory acuity. *Ann. Otol., Rhinol. Laryngol.*, **79**, 95–104.

Strauss, J. S. and Ebling, F. J. (1970), Control and function of skin glands in mammals, In: *Memoirs of the Society of Endocrinologists*, No. 18, Cambridge University Press, Cambridge, pp. 341–71.

Suthers, R. A. (1970), Vision olfaction Taste, In: *Biology of Bats*, Vol. II (ed. W. A. Wimsatt), Academic Press, London, pp. 265–309.

Sutter, E. (1946), Das Abwehrverhalten nestjunger Widehopfe. *Der Ornithologische Beobachter*, **43**, 72–81.

Suzuki, N. and Tucker, D. (1977), Amino acids as olfactory stimuli in freshwater catfish, *Ictalurus catus* (Linn). *Comp. Biochem. Physiol.*, **40A**, 399–404.

Sveredenko, P. A. (1954), Finding of food in the earth by rodents and their conditioned reflexes as to non-food odors. *Zoologischeskii zhurnal*, **33**, 876–88.

Swingle, H. S. (1953), A repressive factor controlling reproduction in fishes. *Proceedings of the eighth Pacific Science Congress*, IIIA, 865–70.

Tavolga, W. N. (1956), Visual, chemical and sound stimuli in the sex discriminatory behavior of the gobiid fish, *Bathygobius soporator*. *Zoologica*, **41**, 49–64.

Tavolga, W. N. (1976), Chemical stimuli in reproductive behavior in fish: communication. *Experientia*, **32**, 1093–5.

Teichmann, H. (1957), Das Riechvermogen des Aales (*Anguilla anguilla* L.). *Naturwissenschaften*, **44**, 242–6.

Teichmann, H. (1959), Concerning the power of the olfactory sense of the eel, *Anguilla anguilla* (L.). *Z. vergleichende Physiologie*, **42**, 206–54.

Tembrock, G. (1968), Communication in selected groups: Land mammals, In: *Animal Communication, Techniques of Study and Results of Research*. (ed. T. A. Sebeoc), Indiana University Press, Bloomington, Ind., pp. 338–405.

Tester, A. L. (1963a), The role of olfaction in shark predation. *Pacific Sci*, **17**, 145–70.

Tester, A. L. (1963b), Olfaction, gustation and the common chemical sense in sharks, In: *Sharks and Survival* (ed. P. W. Gilbert), D. C. Heath & Co., Boston, Mass., pp. 255–81.

Thielcke, G. (1969), Geographic variation in bird vocalisations, In: *Bird Vocalisations* (ed. R. A. Hinde), Cambridge University Press.

Thiessen, D. D. (1968), The roots of territorial marking in the Mongolian gerbil: a problem of the species common topography. *Behav. Res., Methods and Instruction*, **1**, 70–6.

Thiessen, D. D. (1977), Thermoenergetics and the evolution of pheromone communication, In: *Prog. Psychobiol. physiol. Psychol.*, **7** (ed. J. M. Spraguet and A. N. Epstein), Academic Press, New York, pp. 92–191.

Thiessen, D. D. and Rice, M. (1976), Mammalian scent gland marking and social behavior. *Psychol. Bull.*, **83**, 505–39.

Thiessen, D. D., Regnier, F. E., Rice, M., Goodwin, M., Isaacks, N. and Lawson, N. (1974), Identification of a ventral scent marking pheromone in the male Mongolian gerbil (*Meriones unguiculatus*). *Science*, **184**, 83–5.

Timms, A. M. and Kleerekoper, H. (1972), The locomotor responses of male *Ictalurus punctatus*, the channel catfish, to a pheromone released by the ripe female of the species. *Trans. Am. Fish. Soc.*, **101**, 302–10.

Todd, J. H., Atema, J. and Bardach, J. E. (1967), Chemical communication in the social behavior of a fish, the Yellow Bullhead, *Ictalurus natalis*. *Science*, **158**, 672–3.

Tomkins, T. and Bryant, M. J. (1974), Oestrous behaviour of the ewe and the influence of treatment with progestagen. *J. Reproduction and Fertility*, **41**, 121–32.

Trammel, K. (1976), Use and field application of pheromones in orchard pest management programs. *Env. Quality and Safety*, **5**, 69–72.

Tucker, D. (1963), Olfactory, vomeronasal and trigeminal responses to odorants, In: *Proc. 1st Int. Symp. Olfaction and Taste* (ed. Y. Zotterman), pp. 45–69.

Tucker, D. and Suzuki, N. (1972), Olfactory responses to Schreckstoff of catfish, In: *Olfaction and Taste. Proc. 4th Int. Symp. Olfaction and Taste* (ed. D. Schneider), Wissenschaftl. Verlagsges. mbH. Stuttgart, pp. 121–7.

Turner, D. C. (1975), *The Vampire Bat*, Johns Hopkins University Press, Baltimore, Md.

Twitty, V. C. (1955), Field experiments on the biology and genetic relationships of the Californian species of *Triturus*. *J. exp. Zool.*, **129**, 129–48.

Twitty, V. C., Grant D. L. and Anderson, O. (1966), Course and timing of the homing migration in the newt, *Taricha rivularis*. *Proc. natn. Acad. Sci. U.S.A.*, **56**, 864–71.

Twitty, V. C., Grant, D. L. and Anderson, O. (1967), Initial homeward orientation after long distance displacement in the newt, *Taricha rivularis*. *Proc. natn. Acad. Sci. U.S.A.*, **57**, 342–8.

Ueda, K., Mara, T. J. and Gorbman, A. (1967), Electroencephalographic studies on olfactory discrimination in adult spawning salmon. *Comp. Biochem. Physiol.*, **21**, 133–43.

Valenta, J. G. and Rigby, M. K. (1968), Discrimination of the odor of stressed rats. *Science*, **161**, 599–601.

Verberne, G. and de Boer, J. (1976), Chemocommunication among domestic cats, mediated by the olfactory and vomeronasal senses. I. Chemocommunication. *Z. Tierpsychologie*, **42**, 86–109.

Verberne, G. (1976), Chemocommunication among domestic cats mediated by the olfactory and vomeronasal senses. II. The relation between the function of Jacobson's organ (Vomeronasal organ) and Flehmen behaviour. *Z. Tierpsychologie*, **42**, 113–28.

Visser, J. (1966), Colour changes in *Leptotyphlops scutifrons* (Peters) and notes on its defensive behaviour. *Zoologica africana*, **2**, 123–5.

De Vries, H. and Stuiver H. (1961), The absolute sensitivity of the human of smell, In: *Sensory Communication: a Symposium* (ed. W. A. Rosenblitt), MIT Press, Wiley, New York, pp. 159–67.

Vrtiš, V. (1930), Glandular organ on the flanks of the water rat, their development and changes during the breeding season. *Biologicke Spisy Vysoke Školy Zvěrolekařské*, **9**, 1–51.

Walker, D. R. G. (1972), Observations on a collection of weasels (*Mustela nivalis*) from estates in south-west Hertfordshire. *J. Zool.*, **166**, 474–80.

Watkins, J. F., Gehlbach, F. R. and Kroll, J. C. (1969), Attractant-repellent secretions of blind snakes (*Leptotyphlops dulcis*) and their army ant prey (*Neivamyrmex nigrescans*). *Ecology*, **50**, 1098–102.

Watson, J. B. (1910), Further data on the homing sense of noddy and sooty terns. *Science*, **32**, 470–3.

Watson, R. H. and Radford, H. M. (1960), Influence of rams on the onset of oestrus in Merino ewes in the spring. *Aust. J. agric. Res.*, **2**, 65–71.

Welker, W. I. (1964), Analysis of sniffing of the albino rat. *Behaviour*, **22**, 223–44.

Wenzel, B. (1948), Techniques in olfactometry. *Psychol. Bull.*, **45**, 231–46.

Wenzel, B. M. (1968), The olfactory prowess of the kiwi. *Nature*, **220**, 1133–4.

Wheeler, J. W., von Endt, D. W. and Wemmer, C. (1975), 5-Thiomethylpentane-2, 3-dione. A unique natural product from the striped hyaena. *J. Am. chem. Soc.*, **97**, 441–2.

Whitten, W. K. (1956), Modifications of the oestrous cycle of the mouse by external stimuli associated with the male. *J. Endocrinol.*, **13**, 399–404.

Whitten, W. K. (1956), The effect of removal of the olfactory bulbs on the gonads of mice. *J. Endocrinol.*, **14**, 160–3.

Whitten, W. K. and Bronson, F. H. (1970), The role of pheromones in mammalian reproduction, In: *Advances in Chemoreception I. Communication by Chemical Signals* (ed. J. W. Johnston *et al.*), Appleton-Century-Crofts, New York, pp. 309–325.

Whitten, W. K., Bronson, F. H. and Greenstein, J. A. (1968), Estrus-inducing pheromone of male mice: transport by movement of air. *Science*, **161**, 584–5.

Wilde, W. S. (1938), The role of Jacobson's organ in the feeding reaction of the common garter snake, *Thamnophis sirtalis sirtalis* (Linn.). *J. exp. Zool.*, **77**, 445–65.

Williams, G. C. (1964), Measurement of consociation among fishes and comments on the evolution of schooling. *Publ. Mus. Michigan State Univ.: Biol. Series*, **2**, 349–84.

Wilson, E. O. (1971), *The Insect Societies*, Harvard University Press, Cambridge, Mass.

Wilson, E. O. and Bossert, W. H. (1963), Chemical communication among animals. *Rec. Prog. Hormone Res.*, **19**, 673–716.

Winans, S. S. and Scalia, F. (1970), Amygdaloid nucleus: New afferent input from the vomeronasal organ. *Science*, **170**, 330–2.

Winokur, R. M. and Legler, J. M. (1975), Chelonian mental glands. *J. Morphol.*, **147**, 275–92.

Wrede, W. (1932), Versuche über den Artduft der Elritzen. *Z. verlgleichende Physiol.*, **17**, 510–9.

Wright, C. W. and Bourquin, O. (1977), Use of repellent to control crop damage by antelope. *The Lammergeyer*, **23**, 36–9.

Wright, R. H. (1964), *The Science of Smell.* George Allen and Unwin, London.

Wright, R. H. (1976), Odour and molecular vibration: a possible membrane interaction mechanism? *Chemical Senses and Flavor*, **2**, 203–7.

Wright, R. H. (1977), Odor and molecular vibration: neural coding of olfactory information. *J. theor. Biol.*, **64**, 473–502.

Wright, R. H. and Burgess, R. E. (1975), Molecular coding of olfactory specificity. *Can. J. Zool.*, **53**, 1247–53.

Würdinger, I. (1979), Olfaction and feeding behaviour in juvenile geese (*Anser a. anser* and *Anser domesticus*). *Z. Tierpsychologie*, **49**, 132–35.

Wynne-Edwards, V. C. (1962), *Animal Dispersion in Relation to Social Behaviour.* Oliver & Boyd, Edinburgh.

Yarger, R. G., Smith, A. B., Preti, G. and Epple, G. (1977), The major volatile constituents of the scent mark of a South American primate *Sanguinus fuscicollis*, Callithricidae. *J. chem. Ecol.*, **3**, 45–56.

Yu, M. L. and Perlmutter, A. (1970), Growth inhibiting factors in the zebra fish (*Brachydanio rerio*) and the blue gourami (*Trichogaster trichopterus*). *Growth*, **34**, 153–75.

Zarrow, M. X., Gandelman, R. and Denenberg, V. H. (1971), Lack of nest building and maternal behavior in the mouse following olfactory bulb removal. *Hormones and Behavior*, **2**, 227–38.

Zinkevich, E. P., Vitt, S. V., Nikitina, S. B., Kadentsov, V. I., Chizhov, O. S., Tishchenko, A. I., Rosinov, B. V. and Sokolov, V. E. (1973), On the chemical nature of odorous secretions from the "musk" gland of the desman (*Desmana moschata* L.), *3rd Soviet-Indian Symp. Chem. Natural Compounds*, pp. 65–6, publ. "FAN" Tashkent.

Subject Index

(References to figures and tables are in italics)

Taxonomic Index

(References to figures and tables are in italics)

Fish

Amphibia

Reptiles

Birds

Mammals